# GOLD

Susan La Niece

# GOLD

THE BRITISH MUSEUM PRESS

To my parents

*(frontispiece)* Gold oak wreath with a bee and two
cicadas, said to be from a tomb on the Dardanelles,
Greece, 350–300 BC. Diam. (as restored) around 23 cm.

*(right)* Plaque of mythical beast from the Oxus Treasure,
Achaemenid Persia, 5th/4th century BC. L. 6.15 cm.

Susan La Niece has asserted the moral right to be
identified as the author of this work

First published in 2009 by The British Museum Press
A division of The British Museum Company Ltd
38 Russell Square, London WC1B 3QQ
www.britishmuseum.org

A catalogue record for this book is available from the British Library

ISBN   978-0-7141-5076-5

Designed and typeset in Garamond by Redloh Designs
Printed and bound in Singapore by Tien Wah Press

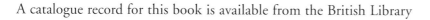

The papers used in this book are natural, renewable and recyclable products
and the manufacturing processes are expected to conform to the
environmental regulations of the country of origin.

# Contents

# Acknowledgements

I am most grateful to the many colleagues whom I have consulted during the writing of this book and I would particularly like to thank Barrie Cook, Paul Craddock, Megan Gooch, Marilyn Hockey, Catherine Johns, Jody Joy, Colin McEwan, Nigel Meeks, Carol Michaelson, Jack Ogden, Venetia Porter and Rachel Ward. Most of all I would like to thank my editor, Nina Shandloff, and my husband, Jeremy La Niece.

(opposite) Bowl from the Oxus Treasure, Achaemenid Persia, 5th/4th century BC. Diam. 12 cm.

# 1 Fact and fantasy

'*Quid non mortalia pectoral cogis, Auri sacra fames!*'
'What do you not drive human hearts into, cursed craving for gold!'
Virgil, *Aeneid* iii.56, around 29–19 BC

We shall never know exactly where or when a bright gold nugget glinting among the pebbles at the bottom of a stream first caught the eye of man (or woman, or child), but no gold has yet been found at the very earliest sites of human habitation in the Near East where copper and coloured minerals were already being collected and treasured, nor in the first primitive societies of Australasia, North America or sub-Saharan Africa.

Gold is too soft for practical purposes, making it far less useful than stone and copper for fashioning effective, durable tools and weapons, so perhaps it is not surprising if it was ignored until societies progressed to a stage where the links between status and display were well developed.

The earliest well-dated gold artefacts to be found in quantity come from burial grounds around Lake Varna near the Black Sea coast of eastern Bulgaria. At that important site, at about 4250 to 4000 BC, gold ornaments were being sewn on to the shrouds of the dead, and skeletons found there were adorned with discs, pendants and bangles. Gold dust was even used to decorate pots. The four richest graves contained some 2,200 gold objects. Intriguingly, large numbers of gold ornaments were also placed alongside copper and other materials in empty graves where there was no evidence that there ever had been a body. These graves have been interpreted as cenotaphs for high-status people who had died far from home. Three of these empty graves contained life-sized human faces, presumably substituting for the deceased, modelled in unfired clay and embellished with gold. This lavish use of gold to honour the dead

Gold strainer for beer or wine from Ur, Mesopotamia (southern Iraq), about 2600–2400 BC. This strainer is a good example of the most common gold-working technique of this time, which was beating it into sheets and cutting to shape. Diam. 12 cm.

Early Bronze Age gold disc from the Isle of Man, Britain. The disc is simply decorated and pierced with two holes, perhaps to attach it to a garment.

Star pendant, from Tell el-ᶜAjjul, Israel, around 1750–1550 BC. This pendant was part of a hoard of jewellery found in southern Canaan. W. 2.06 cm.

gives a clear indication of the significance of this metal to the peoples of Lake Varna at this early date.

Another early find of gold came from a cave at Nahal Qanah in the southern Levant, where eight ring-shaped ingots weighing a total of one kilogram were found close to human bones and the skull of a child dated to the Chalcolithic (Copper Age) period, about 4000 BC. Nearby were rare and important objects including stone maceheads, a large decorated turquoise bead, an ivory point bound with copper and decorated stone bowls. The finds from both Varna and Nahal Qanah show some skill in working with gold, which suggests that somewhere there were as yet undiscovered people who had already started to experiment with this new material.

Gold nugget.

## Occurrence and exploitation of gold

Metals are found in the Earth's crust as metallic grains and nuggets – that is in their 'native' state – or else as minerals (ores), which look more like coloured rock than metal. These minerals have to be processed mechanically, usually by crushing and washing to separate out the metal-rich particles that form just a small part of the ore, and then chemically reduced to turn them to metal. Gold is unusual in being normally found in metallic form, but that does not necessarily make it easier to collect in any quantity. Gold is present as fine particles in rocks formed throughout the Earth's history, from the early Precambrian era, more than three billion years ago, to the present, but in very few areas are there large enough concentrations to be worth exploiting. Gold is even present in solution in seawater; it has been calculated that all the oceans of the world altogether contain over 20 million tonnes of gold, but the cost of extraction is so great that it has never been considered commercially worthwhile. Natural erosion of gold-bearing rocks over many hundreds of years has provided a more accessible source. Constant erosion by sun, snow, wind and rain has caused the tiny gold grains to be washed into streams and rivers where, because of their weight, the current does not have the power to carry them further and they collect at the bottom of pools. Concentrations in river gravels of this waterborne alluvial gold are known as placer gold and it was these placer deposits that were exploited in ancient

20th-century gold panning on the mine waste-tips of the Kolar goldfields of India.

'Goldfields' brooch, Australia, 1855–65, depicting the tools of the Australian goldminer. The action of the water and gravel occasionally rolled small grains together into clumps. These nuggets would have been the most visible indication of the presence of gold, but gold grains, even when as fine as dust, can be collected by exploiting the density of gold. Gold will usually sink to the bottom and can be separated out of the river gravels by washing away the lighter materials. A wide, shallow pan is the best known and simplest tool for collecting gold, and various types of shovel and sluice have been described in the literature from ancient times to the present day. W. approx. 7 cm.

times. For example, in Egypt alluvial gold was washed into the valleys of the Eastern Desert by flash floods and it was this that was extracted in the Predynastic period. The Romans also exploited alluvial gold, recording that gold production from placer deposits in Spain was 200,000 oz a year, a not inconsiderable quantity.

The Greek legend of the quest of Jason and the Argonauts, to the coast of a land far to the east, to find a Golden Fleece, may well have its origins in tales of travellers to gold-rich rivers: the fleece of a sheep is a perfect filter for collecting fine particles of gold from water. The Greek geographer and historian Strabo (around 64 BC–around AD 23) wrote that 'around 400 BC in the country of Saones … the winter torrents brought down gold, which the barbarians collected in troughs pierced with holes and lined with fleeces'. Indeed, until recent times in the Svaneti region of north-west Georgia, people still collected gold from streams in the fleece of a sheep, which was then hung up to dry and the particles shaken and combed out of the wool. The legend of King Midas of Lydia may also have evolved as a fanciful explanation for the actual richness of the River Pactolus in alluvial gold, though by Strabo's time the gold had been exhausted. Legend tells that the greedy

Painted enamel dish showing a scene from the legend of the Golden Fleece, Limoges, France, around 1568. Diam. 23.5 cm.

Midas foolishly wished that all he touched would be turned to gold, with the inevitable consequences when he tried to eat his dinner. According to the tale he prayed to the god Dionysus to save him from starvation. He was instructed to bathe in the River Pactolus, which did indeed cure him and henceforward gold was to be found in the river gravels.

According to the Spanish, most of the gold extracted in South and Central America was obtained from the rivers. One source of AD 1513 described the process, which differs slightly from the panning of river gravels more familiar from Europe and North America: '... they wait until the rivers rise ... and after the floods have passed, and they become dry again, the gold is exposed, having been washed from the banks and carried from the mountains in very large nuggets ...'. This silt was dug out with sticks with fire-hardened points, the earth was sluiced and then panned in the river in shallow oval bowls to separate out the gold. Another strategy of the native South American peoples was to divert streams to wash gold from the river terraces.

As well as alluvial gold, primary sources of metallic gold in quartz veins (reef gold) were certainly mined in antiquity, but later mine workings have often removed much of the evidence. The sixteenth-century text *De Re Metallica* by the German writer Agricola is the earliest definitive text on how to find and exploit gold and other metals, but there are earlier sources which may refer to gold mining. There is an early second-millennium BC text from Ur (in modern-day Iraq): 'red gold. 10 shekels from pounded rock', which must surely refer to mined gold, but where it was mined is not known. The most ancient geological

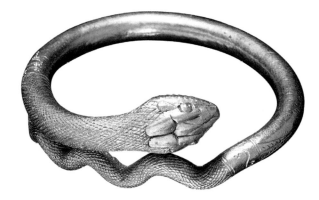

map known, 'La carte des mines d'or' in the Turin Museum, is an Egyptian papyrus. Fragments of it depict a region of Egypt where gold was being mined at about the time of the pharaoh Seti I (around 1320 BC). On it are shown gold mines, roads, miners' houses, quarries and gold-rich mountains, though the exact position of the site shown on the map is problematical.

The Roman author Pliny the Elder (who was killed by the eruption of Mount Vesuvius on 25 August AD 79) recorded what seem to be eye-witness accounts of gold mining. Also from the Roman period there is archaeological evidence for gold mining at Dolaucothi in central Wales and we also know of gold mining in southern India at a similar date.

(opposite top) Thin sheet gold ornament of the Nasca culture of Peru. W. 13 cm.

(opposite below) Two hollow gold armlets. Egyptian, 18th Dynasty (about 1500–1320 BC). Diam. 11.9 cm.

Roman gold bracelet in the form of a coiled snake, excavated at the town of Pompeii which was destroyed by the eruption of Mount Vesuvius in AD 79. Diam. 8 cm.

Sheet gold chest ornament with embossed design of a lizard or frog-man surrounded by birds, from Colombia. Diam. 26.8 cm.

In South America, particularly in Colombia, small mine workings were documented by the sixteenth-century Spanish conquistadors. The native Colombians used stone tools and dug very narrow shafts to reach the gold-bearing quartz veins, but they do not seem to have attempted to extend these shafts along the veins to make underground galleries.

Until the seventeenth century AD, when explosives began to be used in mining, a gold-bearing quartz vein would be broken up by fire-setting – building a fire against the rock face to cause cracking, then further breaking it up with hammers and picks. The pieces were ground up in mortars and washed to separate out the gold particles. This was hard labour, and at state-organized mines it was often carried out by slaves and convicted criminals.

Mercury was also used to extract fine gold particles from crushed rock as it will form a liquid alloy (amalgam) with the gold, leaving the rock behind. This technique was certainly in use in the Islamic world by the tenth century AD and continued in use until recent times, for example in Latin America, often leading to severe pollution of the local environment.

## Alchemy, myth and legend

Gold was marked out from other metals, not just for its value as an indicator of status and wealth but because of its apparent immunity from decay and rust. It was this property in particular which drew alchemists to invest much fruitless effort in the search for the philosopher's stone, a legendary substance that was believed to be the tool to convert base metals such as lead into gold and to create the elixir of life. Twelfth-century European alchemy has its roots in a mix of mysticism, scientific experimentation, astrology and medicine, all drawn from much earlier traditions of both East and West. In the West, alchemy can be traced to Egypt, where adepts first developed it as an early form of chemistry and metallurgy, using their art to make alloys, dyes, perfumes and to embalm the dead. It is widely known that the so-called alchemical transmutation of base metal to gold was used to defraud the gullible, but the true significance of gold to the alchemists was its immunity to the decay which led inexorably to death. Paracelsus, the sixteenth-century physician and alchemist, expressed it as follows: 'When the Philosophers speak of gold and silver, from which they extract their matter, are we to suppose that they refer to the vulgar gold and silver? By no means; vulgar silver and gold are dead, while those of the Philosophers are full of life.'

Gold has been inextricably linked to myth and magic throughout history. The dangers of mining perhaps gave rise to the many legends of fire-breathing dragons guarding piles of treasure deep inside mountain caverns. The Greek historian Herodotus, writing in the fifth century BC, told of a land of one-eyed men and gold-guarding griffins, probably in the gold-rich regions east of the Crimea. One of the strangest legends concerning gold sources, also recorded by Herodotus, is that there were huge, hairy, gold-digging ants in a desert region to the north of India. He tells how these ants, which were bigger than foxes, dug up gold-rich sand in making their burrows. Local

Pottery oil flask decorated with two griffins guarding a pile of gold, Greece, around 380 BC. A griffin is a mythical creature with the head and wings of an eagle and a body like a lion. H. 9 cm.

(opposite) Jewelled and enamelled gold hat badge depicting the Judgement of Paris, Germany, 16th century. The scene shows the moment when Paris, seated and dressed in armour, presents the golden apple to a white-enamelled Venus, with the two rejected goddesses, Juno and Minerva, standing behind her. Diam. 3.6 cm.

(opposite right) Gold scabbard containing a sword with a gold hilt, Sasanian, 7th century AD. Said to come from Dailaman, north-west Iran. L. 196.5 cm.

people came and filled bags with this sand at noon when the ants were sheltering from the sun, and made their escape on fast camels, hotly pursued by the ants which devoured any men or camels they caught. Many explanations have been offered for this tale, perhaps the most plausible being that the ants are a mistranslation of marmots, which dug out deep burrows and could perhaps have accidentally brought gold to the surface. It would be hard to claim that a marmot could pursue and eat a camel, but like many legends, it is probably a confusion of several different stories which were improved and elaborated with the telling, but often had just a grain of truth.

Fairytales and legends from all periods frequently feature something that is golden: princesses are golden-haired, a golden bird brings news and the valiant knight brandishes a golden sword. In these tales gold is sometimes not only a thing of beauty, but also the cause

Gold wire jewellery, late Bronze Age (11th–9th century BC), from Hungary.

of strife or disaster. The golden apple which Paris, prince of Troy, awarded to Venus, goddess of love, led inexorably to the tragedies of the Trojan War and the deaths of many heroes. The history of gold and its exploitation contains many such episodes: the brutal conquest of the native peoples of South and Central America by Europeans greedy for gold, and the hardships and frequent disappointments endured by prospectors of the Californian gold rush are but two examples. Throughout the history of the human race, gold has been a mixed blessing.

# 2

# Status and power

Gold cup, Ur (southern Iraq), 2600–2400 BC. This cup was one of four found on the floor of the burial pit of the 'Queen's Grave' in the Royal Cemetery at Ur alongside human sacrificial victims. H. 12.4 cm.

'Upon thy right hand did stand the queen in a vesture of gold.'
Prayer Book 1662, 45:10

It is no coincidence that at the very time when gold first appears in the cultural remains of prehistoric societies, social hierarchies and elite classes are also coming to the fore. Gold does not have the properties needed for tools and weapons, but it is ideally suited to display because of its colour, brilliance and perhaps most importantly, its ability to resist the corrosion that tarnishes other metals. The Chalcolithic (Copper Age) burials at Varna in Bulgaria are the earliest so far known to contain gold in quantity, and it is the personal items chosen to accompany the dead on the journey to the next world which tell us something of the social status in life of the occupants of the graves. Only a small proportion of the dead were buried with gold ornaments and among those, the archaeologists found just a select few who had really lavish quantities. It is possible to deduce from this

that a small elite class had already begun to dominate and monopolize the gold resources of the community. Similar trends can be seen from the burials of the early Indus civilizations of the Indian subcontinent and in Mesopotamia (modern-day Iraq). At about 2500 BC the ruling elite in the city of Ur went to their final resting place surrounded by their wealth as well as with the bodies of their courtiers. The status in life of the deceased is reflected in the quantity and richness of the objects they were able to take with them into the afterlife.

In Egypt too, gold had become the symbol of wealth and social rank. Prominent chest ornaments known as pectorals first appear in royal burials of the Middle Kingdom (around 2040–1750 BC), but even earlier were depicted in wall paintings of the Old Kingdom (around 2613–2160 BC). Pectorals were worn during life as well as being placed on the chest of a mummy and their decoration was often associated with kingship, and with the protection of the gods. No fewer than twenty-six pectorals were found in the tomb of Tutankhamun.

Gold pectoral in the form of a falcon, inlaid with glass. Egypt, Late Period, after 600 BC. W. 14.8 cm.

In ancient Greece too gold had a status value. In the *Iliad* and the *Odyssey* the poet Homer (around 1000 BC) wrote of gold both as a sign of wealth among mortals and as a symbol of the omnipotence of the immortals.

## China

It must be said that not all powerful elites have chosen gold as the material through which to display their status. In most of ancient China, gold and silver were not valued as highly as they have been in many other cultures and China never had a major gold coinage. The Shang (around 1500–1550 BC) was a powerful warrior elite for whom ancestor worship played an important role. They developed bronze casting to a high level, particularly for the manufacture of vessels for ritual offerings to the ancestors, and they valued jade above all other materials. Gold featured rarely among the treasures buried with the aristocratic elite, though it was sometimes used for decorative purposes. There is a little-known exception to the general rule that gold was not used at this early period in China. Contemporary with the Shang dynasty the people of the city of Sanxingdui, in Sichuan province, made fairly extensive use of gold. Gold is found in their tombs but very little else is known of these peoples. Even though they must have had contact with the Shang, they are not mentioned in contemporary or later writings. Some centuries later, from the Eastern Zhou period (771–221 BC), gold began to be increasingly used in China, though never replacing jade and bronze as the most favoured funerary items.

Later still, in the Tang period (AD 618–906), we know that some senior officials were entitled under the imperial regulations to wear belt sets made in gold, but gold was still further down the ranking than jade.

Cast openwork gold dagger hilt, China, Eastern Zhou period. H. 9.8 cm.

Openwork gold plaque inlaid with semi-precious stones, China, Ming dynasty, 15th century. This plaque is one of a pair that would have been sewn on to a robe. Only the emperor was allowed to use items decorated with five-clawed dragons. W. 18 cm.

## The rod of authority

Rods or sceptres have a long history as symbols of authority, both royal and religious. In the British Parliament a large decorative rod termed the Mace is the symbol of Royal Authority, representing the power and authority that the monarchy has delegated to the House of Commons. A much more delicate rod, from an Etruscan tomb at Taranto, southern Italy, dating to 350–320 BC, is thought to have a religious rather than a royal connection, possibly belonging to a priestess of the temple of Hera. The core of the shaft is now lost – perhaps it was of wood – but the delicate gold sleeve consisting of a fine net of gold wire with tiny cells for coloured enamel has been reconstructed as being of just over 50 cm in length.

(far left) Detail of the top of a delicate sceptre which may have belonged to a priestess of the Greek temple of Hera at Taranto, around 350–320 BC. Diam. (of shaft) 1.2 cm.

(left) Sceptre decorated with bands of gold, shell, lapis lazuli and red limestone, from one of the largest graves in the Royal Cemetery at Ur (southern Iraq), around 2600–2400 BC. L. 41.5 cm.

Gold Mixtec pendant of a nobleman holding a serpent staff prominently in his right hand and a shield in his left, Tehuantepec, Oaxaca, Mexico, AD 900–1521. H. 8 cm.

A rod or sceptre was also a recognized symbol of power in the Americas. We are told by Bernal Diaz, who fought with Cortes in Mexico in the 1519 campaign and observed the last days of the Aztec emperor Moctezuma, that he held a gold and wooden rod when riding in his litter, and that two sceptres were carried before him to warn the people of his approach.

## Kings and crowns

Grand headgear of all shapes and sizes is a long-established means of advertising the importance of the wearer. Tiaras were spoken of by the ancient Greeks, particularly in relation to the rulers of Persia and other eastern lands, but by no means are all crowns made of gold. The crown still firmly set on the skull of an Iron Age warrior buried with his sword and shield at Deal in south-east England is made from two sheets of bronze held together with rivets, and the crowns of the Yoruba chieftains of Nigeria were crafted from beads. The victor's crown awarded at the great games of ancient Greece was of laurel leaves, and flowers make up the traditional crown of the queen of the May at springtide festivals in Britain.

Gold diadems are found in rich graves of many cultures but in many cases they were probably made especially for the funeral of the wearer, presumably to ensure no-one would mistake their social status in the afterlife. The decorated funerary diadem from Hala Sultan Tekké, Larnaca, Cyprus, was found adorning the forehead of the deceased in a tomb dating to about 1400–1200 BC. The decoration on the thin sheet-gold strip was made by stamping with a set of punches of shields and bulls' heads in a repeating design.

Gold burial diadem, Hala Sultan Tekké, Larnaca, Cyprus, dating to about 1400–1200 BC. L. 21 cm.

Gold crowns have been the symbol of royalty in many parts of the world. The crown taken by David from the king of Ammon at Rabbah and used as the state crown of Judah was, according to the second book of Samuel, of gold with precious stones. In England too, the crown of the kings and queens plays a major part in the coronation of a new monarch. The climax of the coronation ceremony comes when the Archbishop of Canterbury places St Edward's Crown on the sovereign's head. This crown was refurbished for Charles II's coronation in 1661, possibly using gold from Edward the Confessor's crown of 1043. A crown is also worn by the British monarch at other important ceremonial occasions such as the State Opening of the Houses of Parliament.

St Edward's Crown, part of the British Crown Jewels.
Wt. 1255 g.

Album painting of Jahangir, ruler of the Mughal empire in India, holding a crown, AD 1620.
H. 11.2 cm.

Other items of attire can give a strong message concerning the status of the wearer. A unique Bronze Age (around 1900–1600 BC) gold cape was found under a mound, in a stone-lined grave at Bryn yr Ellyllon (the Fairies' or Goblins' Hill) near Mold, north Wales. The cape would have fitted over the shoulders and upper arms of a slender person, possibly a woman. Perforations along the upper and lower edges indicate that it was once attached to a lining, perhaps of textile or leather, which has decayed. It must have severely restricted movement of the arms and shoulders; the wearer of it had ceremonial status by simply remaining motionless. The sheet gold cape was beaten out of a single ingot of gold, then punched with fine decoration of ribs and bosses which may be copying the appearance of a beaded textile cloak.

The Mold gold cape, reconstructed from its fragments, Bryn yr Ellyllon, Mold, Wales, around 1900–1600 BC. Wt. 560 g.

## Torcs, the badge of the warrior class

A very early symbol of status was a metal neck ring, known as a torc. These neck rings are found in many parts of the world, from Persia to the Iberian peninsula, and they are frequently depicted in sculptural representations of warriors. A tiny model in gold of a chariot from the

Oxus treasure carries two figures. On the head of the larger, principal figure is a hood or cap, around the front of which is a flat strip of gold, resembling a diadem, and around his neck is a gold wire torc. The driver wears a similar cap without any headband but also wears a wire torc.

A gold neck ring, as well as bracelets and earrings, adorns a bronze figurine excavated from a cult complex at the third-century BC city of Vani in the Republic of Georgia (known in ancient classical times as Colchis), and from Denmark, the antlered deity on the fine silver cauldron from Gundestrup, probably dating to the first or second century BC, also prominently wears a torc.

Such neck rings were not exclusively made of gold but in the old English saga *Beowulf* the torcs worn by great warriors and kings were always wrought in gold. Torcs were clearly important to the Iron Age peoples of western Europe. Large numbers of them made of bronze, silver and gold have been found. The most famous British torcs were found at Snettisham in East Anglia. At least eleven buried hoards of torcs, ingots and coins have been found. Archaeological excavation established that some of these were stacked on top of each other in shallow pits, and one hoard had been buried in two instalments, an upper stack of silver and bronze on top of a deeper deposit of mainly gold torcs. The great gold torc from Snettisham weighs just over a kilogram and is made up of 64 strands of gold wire twisted into a rope. Casting moulds were made for the decorated terminals and attached to each end of this rope so that the molten gold poured into these moulds

Model in gold of a chariot, part of the Oxus treasure, Persia. Achaemenid, 5th/4th century BC. L. 19.5 cm.

bonded directly to the wires. Most of the hoards were buried about 75 BC, and the entire collection is the largest deposit of gold and silver from the whole of Iron Age Europe. These hoards were found towards the centre of a large enclosure, though unfortunately extensive excavations at the site have as yet failed to establish its purpose.

The great torc from Snettisham, Norfolk, Iron Age, about 75 BC. The decorated loop terminals worn at the front were cast on to the gold wire rope. Diam. 19.7 cm; Wt 1098 g.

## Gifts and tribute

'The king in his generosity unclasped the collar of gold from his neck'
*Beowulf*

Historically, lavish gifts marked out the status of both the giver and recipient. A fragment of an Egyptian wall painting from the tomb of Sebekhotep, a senior treasury official at the Eighteenth Dynasty court of Thutmose IV (around 1400 BC), depicts three men carrying plates of gold nuggets with interlinked rings of gold over their arms. They are probably Nubians, as gold was one of the most important products of Nubia. This is the usual way that gold is represented in Egyptian tomb paintings, though in reality no one man could possibly carry the mass of gold shown here. One of Sebekhotep's responsibilities was to deal with foreign gifts brought to the king and such scenes represented his importance as an official and his relationship with the king. Having these paintings on the walls of his tomb indicates that Sebekhotep hoped to enjoy the privileges of office in death as in life.

Wall painting showing Nubians bringing gifts of gold to the court of Thutmose IV, Thebes, Egypt, 1400 BC. H. 71 cm.

Enamelled gold cup, France, around AD 1390. Known as the Royal Gold Cup, this is said to have been presented as a gift to Charles VI of France in 1391 by Jean, Duc de Berry. It is also mentioned in the accounts of the English royal court.
H. (with cover) 26.3 cm.

## Power and the control of gold

For any material to become a symbol of status it is essential that access to it can be controlled. The control of the sources and the trade routes of gold have been bitterly contested throughout history. In the fifteenth and sixteenth centuries, state-sponsored explorers from Europe vied with each other for access to South American gold.

Sheet gold helmet, early Quimbaya, Colombia, AD 500–600.
Diam. (longest axis) 20.4 cm.

Africa too is rich in gold. Although the indigenous peoples south of the Sahara had little use for it, once foreign powers had realized the abundance of gold there was inevitable competition for control of the wealth. The southern coast of Ghana in West Africa used to be referred to by Europeans as the Gold Coast because of its vast local gold resources. From the end of the seventeenth century this was exploited by the Asante who traded gold and slaves in return for European

Gold coin struck in AH 1040/ AD 1631 by Sharif al-Walid, ruler of Morocco. Diam. 2.9 cm.

firearms with which to extend their rule. On the southern edge of the Sahara are Tombouctou and Gao, which were trading centres for the trains of caravans carrying gold, ivory and slaves and bringing back salt and other goods from the north African kingdoms. A description of Tombouctou dated to 1526 reads: 'Instead of coined money, pure gold nuggets are used; and for small purchases, cowrie shells which have been carried from Persia.' Ahmad al-Mansur of Morocco (died AD 1603) conquered the gold-rich regions of Sudan and was able to control such huge supplies of gold that it was said he paid his officials in pure gold. At the gates of his palace in Marrakesh, '1700 smiths were daily engaged in striking (gold) dinars ...'. This superabundance of gold earned him the honorific al-Dhahabi, 'the Golden'.

In southern Africa, gold from Zimbabwe was traded with the Arabs and later with the Portuguese. The vast city of Great Zimbabwe was built over a long period of time, from around AD 1200 to 1400, and was surrounded by a twenty-metre-high wall. Archaeological evidence found at the site shows that it was the centre of a trading network that stretched across Africa and out to the Indian Ocean ports.

It is not only trade which has been controlled by elites. At various periods and in different cultures, sumptuary laws were made, often unsuccessfully, to forbid the wearing of gold except by certain privileged classes. Early in the Roman Republic, a senator sent on an embassy would receive a gold ring, whereas all other senators were allowed only to wear iron rings. Later the privilege was extended to other classes of citizens until all freeborn men could wear a gold ring, but freedmen were generally only allowed to wear silver and the iron ring became a mark of slavery.

In France, in the early seventeenth century, Louis XIII issued edicts regulating 'Superfluity of Dress' that prohibited anyone but princes and the nobility from wearing gold embroidery or caps, shirts, collars and cuffs embroidered with metallic threads or lace.

Senior chiefs of the Asante of west Africa could have gold ornaments only with the permission of the Asantehene, the King of the Asante, who received taxes in return for the manufacture of the item.

Communities which lived solely by hunting, fishing and gathering wild fruits have little use for gold, but the development of complex societies with regular interchange between neighbouring communities has led to an increasing use of gold as a visible display of status. Gold medals are awarded to the first-place winners at the Olympic Games and gold is still the metal of choice for a wide range of high-status items from Rolex watches to royal regalia.

Delicate casting of three elephants with birds sitting on their backs, Asante, Ghana, 19th century. Possibly either broken from a finger- or toe-ring or intended for mounting on a cap. Diam 4.7 cm.

# 3 Money

'God created the two mineral "stones", gold and silver, as the [measure of] value of all commodities'.

Abdel Rahman ibn Khaldun of Tunis (AD 1332–1406), vol. II, 313

Almost anything can and probably has been used for 'money' as a medium of payment – so why is gold particularly associated with coinage and financial reserves? Why did it become a widely recognized and accepted standard of value? The answers lie partly in its limited availability and also in the properties of gold. Gold does not degrade, it can be stored or hidden without fear that it will rot, rust or be eaten by moths or termites. It is not destroyed by plague, flood, fire or famine. It is relatively portable. It can be melted down or cut into smaller pieces, though it does have limitations when it comes to everyday transactions – the amount of gold needed to pay for a loaf of bread would normally be so small that it would be lost in a purse or pocket even if it could be accurately cut and weighed.

There is one factor which is vital to gold's acceptance as a means of exchange – that its standard of purity can be guaranteed, or at least that the recipient of the gold is confident of its value and that it will be accepted at equal value next time it is traded on. Weight is an easily measured criterion on which to judge value, but it is not infallible: there are many ways that have been used by fraudsters to deceive the unwary.

The first coinage in a form we would recognize, that is standard shaped pieces of metal of specific weight and composition (fineness) and with identification markings stamped on them, was not issued until the late seventh century BC. Long before that, precious metal, particularly silver, was being used as a means of storing and exchanging wealth in Mesopotamia. Three thousand years ago in Egypt, which had access to the rich mining resources of Nubia, gold as well as copper and silver was being hoarded and exchanged by weight. Excavators at the city of el-Amarna, built by the pharaoh Akhenaten (about 1352–1336 BC), found a pot containing a hoard of silver and gold pieces, many with weights approximating to multiples and fractions of the later New Kingdom standard weight system of the deben (91 grams) and the kite (9.1 grams). The pieces were of various shapes and sizes, including bar-shaped ingots which were probably cast into a groove impressed with a stick or finger into sand or clay, and rods of metal bent into a ring shape, perhaps for ease of transport when strung on a cord.

Electrum (an alloy of gold and silver) coins were first struck in the late seventh century BC in Lydia, in what is now western Turkey. These early coins were cast, oval-shaped pellets of an attractive pale golden colour. They are found in different sizes; the heaviest is 17.2 grams, with smaller coins corresponding to regular fractions of the larger ones, down to minute ones less than 0.2 grams in weight. Many of the coins are stamped on one side and have a punch mark on the other. The meaning of the marks is not known and does not seem to relate to their size and weight. Element analysis of this electrum coinage shows it to be of remarkably consistent composition, with about 44% silver and 2% copper in the gold, which is more silver than would be expected of naturally occurring placer gold. This gold, sifted from the river gravels of the Pactolus, like most native gold, was combined with silver in variable proportions (see chapter 1). It is thought that in this period, before there were methods of routinely removing silver from quantities of native gold, silver was deliberately added as a means of ensuring a consistent gold content for these coins. This of course assumes that they were able to assay or test the purity of the gold. The ability to refine and assay precious metal (see chapter 7) was central to the early concept of coinage, based as it was on trust in the value of the metal in the coin.

Following on from these electrum coins, the Lydians produced separate gold and silver issues. This would seem to be the point in history at which a process for separating silver and gold on a large scale was mastered. This high-purity coinage is attributed to Croesus, the last king of Lydia, who ruled from about 561 to 547 BC, and who is still a byword for wealth even today. Scientific study of finds from excavations at his capital at Sardis has produced conclusive evidence that the separation of gold and silver was being carried out there, at what is so far the earliest known gold refinery in the world.

Gold croesid minted in Lydia, Turkey, about 550 BC. Wt. 8 g.

## European coinage

In many of the societies with a monetary system based on precious metals, silver was far more commonly used than gold. The coinage of the Greek city states was almost universally of silver, with rare exceptions to meet financial emergencies, usually precipitated by wars. In 357 BC Philip II of Macedonia seized the gold mines of Mount Pangaeus in Thrace (modern Bulgaria) from the Athenians. With the gold he issued a coinage and financed a standing army to make his conquests of neighbouring territories. Philip's son Alexander the Great (336–323 BC) then consolidated the Greek empire with his conquest of the Persian empire, securing the immense gold treasure built up by the Persians from gold sources on the River Oxus in northern Afghanistan. Alexander is reputed to have taken over 22 metric tonnes (700,000 troy ounces) of gold coins in loot from the Persians.

At some point in the third century BC, the Celtic peoples of continental Europe (from the Balkans through Germany and Switzerland to France and Belgium) began to manufacture copies of the gold coins issued by Philip and Alexander. It has been suggested that northern European mercenaries brought home these gold staters, spreading the idea of coinage to their native lands, even down to the design, with a head on one side and a chariot and horses on the back.

The Roman Republic adopted a silver coinage, similar to the Greek issues, alongside bars and discs of bronze. Gold coinage (aureus) only became a regular part of the Roman monetary system with the establishment of a stable empire under Augustus. In the third century AD the Roman gold coinage was officially dissociated from the silver denomination and was valued at its bullion value. The Roman silver coinage had long since been accorded a face value quite unrelated to the weight of silver it contained, allowing the state to debase their silver content with copper. Breaking the link with silver coinage meant the value of the aureus, weight for weight, was little different from that of any other valuable commodity such as jewellery and plate. The gold coin was used in part to buy the loyalty of the army to the emperor.

In the early fourth century, the emperor Constantine the Great introduced a lighter gold coin – the solidus – which for the first time became the principal element of the Roman precious metal coinage and indeed survived on into the Byzantine empire with barely any change in standards as late as the eleventh century. In the late fourth century AD, there was a shortage of gold in the Roman empire. Taxes in the Late Roman and Byzantine empires were only payable in gold but there were real concerns about the circulation of counterfeit gold coinage. To get round the problem of having to test every coin, the emperors ordered, in about AD 367, all gold coins paid in tax to be melted down into bars for testing, before being turned back into coin. The gold bars illustrated here represent the intermediate form which the gold took after melting and before it was turned back into coins.

Gold bars marked with assayers' stamps, certifying the purity of the metal, Roman, late 4th century AD, found in Egypt and Romania. L. (longest) 18.8 cm.

(above) Gold mancus of Ethelred II of England, minted in Lewes, southern England, AD 1003–16. This is one of the few Anglo-Saxon gold coins to have survived. Is this because few were minted or because they have been recycled? Diam. 2 cm.

(right) Gold and garnet-inlaid pendant cross found at Wilton, Norfolk, England, AD 675–700. In the centre of this Anglo-Saxon cross is a solidus of the Early Byzantine Emperor Heraclius dated to AD 613–30. H. 4.7 cm.

Precious metal continued to be of major significance in the post-Roman world of southern and western Europe, but a centralized monetary system ceased to exist. Gold and silver was stored in the private treasure houses of powerful individuals. In some former Roman provinces, above all Britain, coinage ceased to be officially produced or provided, but coinage still continued in use at some levels and 'antique' gold coins were mounted for use as decorative elements in jewellery.

Outside Italy the coins of the so-called barbarian kingdoms were almost exclusively gold and stocks of gold in circulation dwindled because of the tendency of the Church and secular rulers to hoard it in their treasuries. Furthermore, the payment in gold for eastern luxuries such as spices and fine textiles drained the coffers of the rich. The main coin was the small tremissis, worth a third of the Roman and Byzantine solidus. The gold coinage gradually became debased with silver through the seventh century, and by its end silver coinage dominated, with gold only used for a tiny number of special prestigious coins.

The Byzantine gold coinage originating from the solidus of Constantine the Great remained a stable and trusted trade tool for the Mediterranean world and beyond. With a revival of standards in 1092 under Alexius I Comnenus, it continued to be strong until the fall of Constantinople in the fifteenth century.

After half a millennium of dominance of silver as a coinage metal, in 1251–2 Genoa and Florence began minting gold coins – the genovino and the florin – followed by Venice, which began to produce the ducat in 1284, though still continuing to use Byzantine coins. Much of the gold for these coinage issues was still coming from Africa. The trans-Saharan trade swung westwards away from Egypt to be more accessible to the Italian cities trading with North Africa. The expansion of international trade and the huge expense of wars, including the Hundred Years' War between England and France (1337–1453) and the Crusades, all required large-scale transfers of funds. The availability of a high-value commodity such as gold became a necessity for these transactions. In the later fifteenth century the Portuguese gained direct access to West African gold, bypassing Italian and North African intermediaries. A new coinage system in Spain in 1497 included the issue of the gold excelente, copying the Venetian ducat in size. In the sixteenth century precious metal – gold and, in even larger quantities, silver – flooded into Europe from Central and South America.

Obverse of gold coin of the Abbasid ruler al-Muqtadir, minted AH 314/AD 926. Wt. 2.73 g.

## Islamic gold coinage

The earliest Islamic gold coins, produced in the late 680s, were adaptations of Byzantine gold coinage. Even the name of a coin, dinar, came from the Byzantine denarius aureus. The Byzantine coinage was of course explicitly Christian so images of the cross were removed, but the traditional portraits of rulers were also considered inappropriate and in AD 696 (AH 77), with the coinage reform of the caliph Abd al-Malik and the establishment of a new standard weight for gold of 4.25 grams (also known as the mithqal), these portraits were replaced with religious inscriptions. There were exceptions to this rule, for example the coins of the early nineteenth-century Qajar ruler of Iran, Fath ᶜAli Shah, show him seated on a throne with crown and sabre.

The principal sources of gold for the dinars of the earliest period were Byzantine coins and confiscated treasure. It is thought that some gold was mined in the mountainous region of western Saudi Arabia, but the Muslim expansion into Africa opened up vast new sources of gold and a great quantity of this African gold was minted as dinars. The dinar and the Venetian ducat were the chief tools of Mediterranean trade, though Byzantine gold was still important well into the thirteenth century. In 1425 the Mamluk sultan al-Malik al-Ashraf Barsbay introduced the ashrafi, a gold coin valued on a par with the ducat, weighing 3.41 grams of high-purity gold. The ashrafi became the Islamic gold coin of Iran, the Ottoman lands and India.

## India and South-east Asia

No gold coins were issued in India before the first century AD, though gold was valued as bullion. The great kings of the Gupta dynasty issued a particularly important gold coinage between the fourth and sixth

Gold stater of Kanishka I depicting the god Shiva on the reverse, Kushan, India, around AD 100–125.

centuries AD, but as late as the seventh to eleventh centuries the Palas dynasty ruling in the regions of Bihar and Bengal issued no coinage and used gold dust and cowrie shells as currency, although coinage had been in use in the region previously. The Islamic invasion of 1205 brought silver and gold coinage back to the area.

The Indian tradition of gold and silver coinage spread eastwards and by the thirteenth centuries it had reached as far as present-day

Hoard of coins and scrap gold found on the coast of Thailand. The coins from the hoard date to the ninth century AD. They were found together with a small leaf ornament, gold beads, fragments of wire and other scrap gold, and a series of bosses and flowers made from gold granules. The combination of scrap gold and coins in the hoard suggests that gold in any form was being used as bullion in the maritime trade of the region.

Indonesia and the Philippines. In Java, inscriptions dating to the second half of the ninth century provide the earliest evidence of a currency system in the region based on the use of standardized weights, one of which was known as the masa. The abbreviation mas came to mean 'gold' or 'wealth'.

## China and Japan

China is unusual in having had no major gold currency. According to the *Guanzi* (Book of Master Guan), 'The early kings put a value on things from the furthest distance that were difficult to find. They saw pearls and jade as superior money, gold as medium money and spades and knives [Chinese bronze money] as inferior'.

This comment on the relative value of different materials is attributed to a Chinese minister who died in 645 BC, though the written record was many centuries later. It confirms that gold was not the most valued material in China and that it was not made into standard forms which might be recognized as coinage. On the other hand, spade and knife money of bronze, which was made under licence to a standard design albeit initially in the shape of tools, was used for payment. Marco Polo, writing at the end of the thirteenth century AD, recorded that in southern Sichuan province in western China, gold bars and salt cakes were valued according to their weight for business transactions. Gold and silver western coins have been found in Chinese tombs, but they seem to have represented valued gifts and offerings rather than legal tender in

Gold 10 ryo (ounces) oban, made of hammered gold. Japan, AD 1860 (Man'en Era). The value and signature of the Goto family, hereditary superintendents of the mint, were handwritten in ink. L. 13.4 cm.

commercial transactions. The rare issues of Chinese gold coins were generally intended as presentation pieces and not for circulation.

The first coins made in Japan were silver followed by bronze issues, but in the 1580s gold oban ('large stamped [piece]') began to be made. The feudal lord Toyotomi Hideyoshi (1536/7–98) and wealthy merchants of the Kansai district of central Japan gained a monopoly on Japan's metal mines and began to mint gold oban of fixed quality. The earliest oban were quite plain, but inscriptions and stamp marks were added later to overcome the problem of forgeries.

The New Currency Law of 1897 introduced the yen as a basic unit, worth 750 milligrams of pure gold. Three gold coins were issued: the 20 yen, 10 yen and 5 yen coins, as well as six smaller denominations in silver, cupro-nickel and bronze. The New Currency Law also stated that Bank of Japan notes would be convertible into the new gold coins.

## Africa

The gold resources of Africa had been exploited from at least as early as the second millennium BC, in particular by the Egyptians. Later exploitation and trade of African gold by Arabs and Europeans fed the increasing numbers of gold coins in circulation, but the issue of gold coinage by African states for local use was almost unheard of. A gold, silver and bronze coinage had been issued by the Ethiopian kings of Aksum in the third to seventh centuries AD, but by the early eighteenth century European merchants recorded that there were no coins minted in that region: '... large payments are generally made in Ingots of Gold ... and for small payments Salt Bricks dug out of the mines are adopted'.

Gold krugerrand, Republic of South Africa, 1980. Typical of modern gold bullion pieces, it has no currency denomination and its worth is in its weight, in this case 1 troy ounce of fine gold.

English guinea, so called after the source of the gold, the Guinea Coast of West Africa. Large quantities of African gold were brought to England by the Royal Africa Company, which dealt in gold, ivory and slaves.

Before the nineteenth century, West Africa was one of the world's richest sources of gold, but the local people valued copper and brass and used these as currency for payments. European brass was imported in large quantities, and shells were used as currency in gold-rich Congo in the late sixteenth century.

## The Americas

There is no history of coinage in the Americas until Europeans arrived in the sixteenth century AD. The outstanding gold artefacts produced by the peoples of South and Central America are well known, and this gold was much coveted by the invaders. Although the native peoples clearly appreciated gold, it is evident that their value system differed diametrically from that of Europeans and its appeal to them was intricately bound up with its appearance and symbolic associations rather than as an indicator of wealth. The invaders seized all the gold they could lay their hands on and sent it back to Europe, where much of it was made into coins. In 1692–4 the discovery of rich gold mines at Minas Gerais ('General Mines') in Brazil, which was under Portuguese control, inspired the first gold rush of the Americas. Many thousands of slaves were forced from Africa to work the mines. The output of gold peaked around 1720 with enough to last, it was thought at the time, 'for as long as the world shall endure'. In a couple of decades the world's supplies may have doubled. The gold of Minas Gerais was minted into two series of coins, one intended to circulate in Brazil itself.

On 15 April 1790 the United States House of Representatives directed the Secretary of the Treasury, Alexander Hamilton, to explore the possibilities of establishing a mint. North America had until then relied on imported foreign coins. His extensive report contained some insights into the relative roles of gold and silver in a monetary system at that period: 'As long as gold, either from its intrinsic superiority, as a metal, from its greater rarity, or from the prejudices of mankind, retains so considerable a pre-eminence in value, over silver, as it has hitherto had, a natural consequence of this seems to be that its condition will be more stationary. The revolutions, therefore, which may take place, in the comparative value of gold and silver, will be changes in the state of the latter rather than in the state of the former.'

The coinage established by Alexander Hamilton was based on a gold standard eagle ($10), half eagle and quarter cagle, with silver (dollar and dime) and copper (cent) coins. The source of gold for the

Old 10-dollar coin commonly known as an eagle, minted in Philadelphia, USA 1795. Wt. 17.48 g.

Gold sovereign of George III
with a design of St George and
the dragon by Benedetto
Pistrucci, minted in 1818. The
sovereign replaced the guinea
under the British Coinage Act
of 1816.

eagles was largely from recycled jewellery and foreign coins until the California Gold Rush (1848–55) transformed the supply. US law prohibited states but not individuals from coining money, which is the reason for the variety of privately produced coins of gold from American mines minted in the nineteenth century.

## The gold standard

The monetary system with a fixed price for gold, issuing gold coin, or notes redeemable in gold, became known as 'the gold standard'. This was one answer to the endless problems of a bimetallic system, with two precious metals, gold and silver, whose relative value could shift radically with their availability. This principle of the gold standard can be detected behind many of the early issues of gold coinage worldwide, but it was not formally established until the nineteenth century. Britain went on to an unofficial gold standard in 1717 when Sir Isaac Newton, then Master of the Mint, established a fixed price for gold. Britain adopted a formal gold standard in 1821 at the end of the Napoleonic wars. The rest of Europe, however, remained on a silver standard until the 1870s, when the supply of gold was boosted by the new discoveries from the United States and Australia. Germany switched to gold in 1871, Scandinavia in 1874, The Netherlands in 1875, France and Spain in 1876, Russia in 1893 and Japan in 1897. The United States remained on a bimetallic system of gold and silver until 1900 when the Gold Standard Act confirmed the supremacy of gold. Eventually fifty-nine countries were on the gold standard, with China on silver as the only main exception.

The gold standard was to act as an economic regulator. In the case of a trade deficit, gold would flow out to meet the deficit, and so prices would fall. The fall in prices would lead to favourable conditions for increased exports, leading to a trade surplus and gold would flow back into the country. This system was, in theory, self-

Gold coin, minted 1853, Port Phillip, Victoria, Australia. This coin was privately produced by an English firm intending to strike gold coins for bullion and sell them from their Melbourne store, known as the Kangaroo Office. The back of the coin states that it contains 'fine Australian gold. Two ounces'.

perpetuating, but it was not able to withstand the pressures of a major war. With the beginning of World War I in 1914, the minting of gold coinage largely stopped and bank notes were introduced into regular circulation. In 1933, during the Great Depression, the United States government recalled all gold, including coins, from their citizens. After that, the era of almost universal gold coinage was over.

## Buried treasure

In times of uncertainty, gold bullion has been the security of last resort in many cultures. Much of the treasure which survives today was buried for safety and never retrieved. For example, the Fishpool hoard, comprising 1,237 coins and jewellery, found in Sherwood Forest in England, was probably buried as a consequence of a failed rebellion against the Yorkist king Edward IV in the first decade of the

Jewellery from the hoard found at Fishpool, Nottinghamshire, England. The hoard dates to the Wars of the Roses (1455–85) when the houses of York and Lancaster vied with each other for the crown of England.

Wars of the Roses. The face value of the hoard when deposited was about £400, equivalent to around £300,000 today. The coincidence of the date of the burial of the hoard with the failed rebellion makes it tempting to surmise that it was part of the Lancastrian royal treasury which might have been entrusted to someone fleeing south after the Battle of Hexham on 15 May 1464.

## The value of gold

'The value of the yellow metal, originally chosen as money because it tickled the fancy of savages, is clearly a chancy and irrelevant thing on which to base the value of our money and the stability of our industrial system.' D.H. Robertson, *Money* (1928)

This sprint through the role gold has played in economic history should perhaps be qualified by a look at areas where it has not been so dominant. Gold and gold coins have through history been particularly linked with an ethos of material productivity and profit. Where this ethos did not exist, for example parts of Australasia, Africa and Asia, including places such as the Congo where gold was available as a natural resource, then gold was not much regarded. In these societies transactions were, and in some cases still are, social rather than commercial in nature and relatively local. In the case of fines paid for transgressions, for example, or payments for marriage contracts, gold has no advantage over any other acceptable commodity. It is easy to forget, from a western perspective, that gold has not been universally acclaimed. Nevertheless, in spite of the passing of the gold standard and the ending of the use of gold in daily circulation, the legacy of that metal still has a certain hold. Although the modern pound coin of Great Britain does not contain a trace of gold, it is golden in colour, and British bank notes still bear the legend 'Promise to pay the Bearer on Demand the sum of x pounds' (of gold), wording that dates back to the goldsmiths' notes of the seventeenth century when this promise was intended to be taken literally.

# 4 The goldsmith

I the goldsmith make valuable things
Seals and golden signet rings
Costly pendants and jewels
Set with precious stones
Gold chains, necklaces, bracelets
Goblets and beakers
Also silver dishes and bowls
For whoever is willing to pay me.
Jost Amman, *Book of Trades* (1568)

Historically the goldsmith is an enigmatic figure. Even when the name of a goldsmith is recorded, it is rare to be able to link him (for it seems it was usually a man) to an existing work, as pieces were hardly ever signed. The maker's mark, the unique mark of a goldsmith's workshop stamped on to precious metal items, was not introduced in England until 1363 and even then it was not used

Gold cups for drinking chocolate, made in 1700 for Lady Palmerston. Chocolate was then an expensive luxury with medicinal properties attributed to it. Inscriptions on the insides of the handles and base of the cups read (on one): 'DULCIA NON MERUIT QUI NON GUSTAVIT AMARA' (he has not deserved sweet unless he has tasted bitter) and 'MANIBUS SACRUM' (to the shades of the departed); (on the other) 'Think on yr Friends & Death as the chief' and 'MORTVIS LIBAMVR' (let us drink to the dead). The cups were made from melted down mourning rings, hence the morbid sentiments. H. 6.5 cm.

universally. In the medieval period named goldsmiths are known from wills and inventories but little detail is known of individual goldsmiths, their workshops, status and how much creative control they had over their work.

## The status of the goldsmith: artist or artisan?

Goldsmiths have not always been accorded high status; indeed some were slaves, albeit highly valued. In the medieval period a number appear to have been members of monastic communities. According to an early medieval Irish law tract, the goldsmith had the same honour-price (the payment owed for offences against him) as the coppersmith and the blacksmith, so presumably he was considered to be comparable in status to such artisans of everyday items. On the other hand goldsmiths with influential patrons certainly could prosper and considered themselves to be artists superior to other craftsmen. One such man, who in his autobiography less than modestly proclaimed himself to be a matchless artist, was Benvenuto Cellini (1500–1571), an Italian goldsmith who was incidentally also a painter, sculptor, soldier and musician. He was without doubt something of a braggart, with a volatile temperament, but his work was highly esteemed in his lifetime by the most influential of Renaissance patrons. It was the patronage of cardinals and of the Pope which allowed him to escape punishment for murder on no fewer than two occasions. Some of his works in gold survive, including a sculptural gold and enamel salt container for the dining table. The magnificent gold morse (a clasp for an ecclesiastical cape) set with a huge diamond and made for Pope Clement VII, the competition for which is so graphically described in Cellini's autobiography, was among the plate and jewels paid by Pope Pius VI to Napoleon in 1797. It no longer survives, though a drawing of it was made over a century later.

F. Bertoli del.

(opposite) Gold font made for the baptism of the Duke of Portland's grandson in 1797/8. It was designed by Humphrey Repton (1752–1818) and made by the workshop of Paul Storr (1771–1844), one of the most outstanding and well known of London goldsmiths in the first half of the nineteenth century. He ran an independent firm before joining the Royal Goldsmiths, Rundell, Bridge and Rundell, in 1811. H. 33 cm; Wt. 11 kg.

(left) 'Siren', life-sized cast gold statue by Marc Quinn, AD 2008. This 18-carat gold sculpture, made from casts of the face and body of supermodel Kate Moss, is hollow yet still weighs 50 kg. It is claimed to be the largest gold statue made since ancient Egyptian times.

## Workshops and tools

A goldsmith's workshop leaves little evidence in the archaeological record. Goldsmiths, unlike blacksmiths, glass workers or potters, do not need large hearths or distinctive structures for their work. Gold is far too valuable to be left lying around as waste, though discarded crucibles and moulds may retain traces of precious metal which can be detected by scientific analysis. The tools are not visually easy to distinguish from those of other crafts after their wooden handles have rotted and the iron rusted, and anyway no goldsmith would routinely throw his tools away, though occasionally a cache of goldsmith's tools does come to light, for example at the Viking age settlement of Hedeby on the Danish-German border. Tools found in a grave may hint at the trade of the person buried there. A type of fine-grained stone implement, termed a cushion-stone because of its shape, has been interpreted as a tool for smoothing and shaping sheet gold, therefore implying that the owner was a goldsmith. It has been suggested that the man who was buried with archery equipment close to Stonehenge, at the period of the earliest metalworking, may have worked sheet gold as he was equipped with, among other things, sheet gold ornaments and a cushion-stone. Craftsmen did not necessarily confine their skills to one material only, and there is evidence that many goldsmiths were happy to turn their hands to whatever was wanted. Certainly gold and silver items were and are commonly produced by the same workshop. Additionally the responsibility for the minting of coins was also historically a role of the goldsmith.

Fragment of painted plaster from the tomb of Sebekhotep, Thebes, Egypt, 18th Dynasty, around 1400 BC. This scene depicts a goldsmith's workshop. The men in the upper part are probably stringing beads, while one man in the lower part is making them into the large collar shown on his lap. At one edge a craftsman holds a piece of metalwork in a brazier while he directs the flames with a blowpipe. Above are completed precious metal objects. W. (max.) 66 cm.

The basic tools have changed little over time. A wooden bench and a source of heat for melting metal and soldering, as well as hammers, an anvil or stake, pliers, tweezers and engraving and

Detail of an engraving from the title page of a book published in Nuremberg, Germany, about AD 1690–1730, showing delicate goldsmiths' tools including pliers, tweezers, crucible and a small mortar and pestle. W. 26 cm.

Goldsmith's box, Iran, 19th century AD. Containing three sets of steel balances, with weights and tools, tweezers and measuring spoons, the box is fitted with a bell which sounds when the catch is opened and there is a mirror in the lid so the goldsmith can watch what is happening behind him while he weighs out the gold. L. 64 cm.

tracing tools are some of the basic requirements. Chisels, snips and shears can be used for cutting and clamps and vices for holding the piece. A burnisher (a rounded stone or metal tool) is used to smooth the surface. Casting requires small crucibles in which to melt the gold and tongs to hold the crucibles. Moulds, usually of clay but sometimes other materials such as stone and cuttlefish bone, have been used. There are numerous specialized tools for decorating gold sheet, including punches, dies and doming blocks. Some of the tools in the modern workshop were not available until steel production became reliable, from around the mid-eighteenth century in Europe. Piercing saws must have fine steel blades, so until a supply of steel blades could be relied upon, pierced work was executed with punches and chisels. Quality polished steel rollers were essential for rolling mills which were able to produce more regular thicknesses of metal than was possible by hammering. They were initially developed for coin production and only in the later eighteenth century did they begin to be used in jewellery manufacture, along with die-stamping for mass production.

(left) Front and back views of a pierced section of a gold and garnet necklace, from Silivri, Turkey, 4th century AD. The openwork was not done with a piercing saw, as it would be today. The appearance of the front is quite different to that of the back, where it is obvious that the initial holes were made with a sharp point, hammered through from the front. The holes were then opened out, on the front only, with a small chisel to form the fine tracery. L. 4 cm.

(below) Sheet gold ornament from Ecuador, pre-AD 1500. The thin sheet gold has been worked in high relief over a former to shape the face. Diam. 10 cm.

## Goldsmithing techniques

Some historical handbooks of metalworking techniques have been handed down, for example *De Diversis Artibus*, a manual of painting, glass and metalwork which seems to have been compiled in the early twelfth century by a monk writing in Germany under the pseudonym Theophilus. Most historical descriptions of craft techniques, however, are not original compositions. They have been copied many times and collected together from other manuscripts or oral traditions, changing in the process and thus confusing to follow.

The gold artefacts themselves hold much information about how they were made and scientific study of ancient pieces has played an important role in revealing goldsmithing techniques of

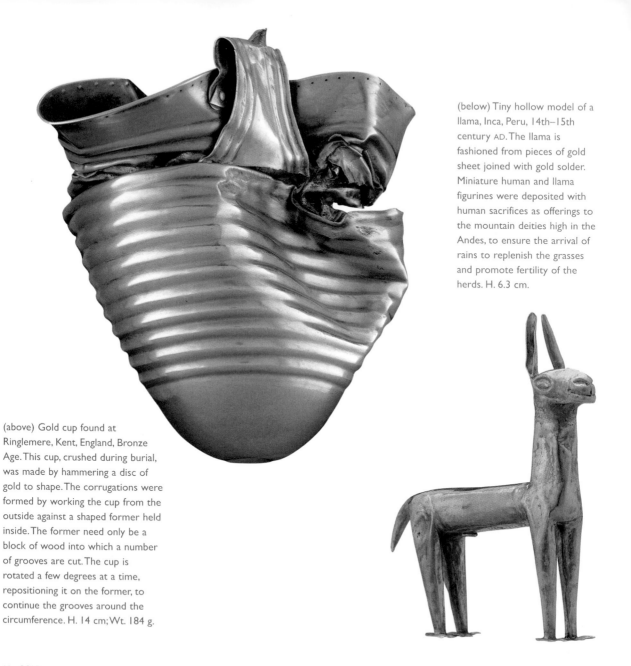

(below) Tiny hollow model of a llama, Inca, Peru, 14th–15th century AD. The llama is fashioned from pieces of gold sheet joined with gold solder. Miniature human and llama figurines were deposited with human sacrifices as offerings to the mountain deities high in the Andes, to ensure the arrival of rains to replenish the grasses and promote fertility of the herds. H. 6.3 cm.

(above) Gold cup found at Ringlemere, Kent, England, Bronze Age. This cup, crushed during burial, was made by hammering a disc of gold to shape. The corrugations were formed by working the cup from the outside against a shaped former held inside. The former need only be a block of wood into which a number of grooves are cut. The cup is rotated a few degrees at a time, repositioning it on the former, to continue the grooves around the circumference. H. 14 cm; Wt. 184 g.

Gold plaque worked in relief from the front (chased) to depict the marriage of Alexander the Great and Roxanna. Made by Ishmail Parbury, 1745, London, England. The plaque is now set as a lid in a 19th-century tortoiseshell snuff box. Parbury is regarded as one of the finest chasers of his time. W. 6.2 cm.

the past. Gold does occur naturally in metallic form, sometimes in sizeable lumps (nuggets) and some very early pieces of goldwork were made simply by hammering a nugget of native gold to shape. The manufacture of most gold items, however, begins with melting the metal and pouring it into a mould, either a simple ingot mould, or a mould in the form of the finished article. The cast ingot is the starting material from which a goldsmith can fashion a huge variety of forms, from thin gold sheet to wire, from a bowl to a ring. A cup, for example, can be made from a flat disc by holding it firmly against a stake and then hammering to force the metal progressively into the form of a hollow container.

Casting molten metal into a mould which is in the form of the desired article has advantages. If the mould is reusable or can be made quickly then casting is a rapid method of manufacture. However, it is wasteful of metal as more must be melted than is needed for the finished article, to allow for shrinkage of the metal on cooling and other losses. Where the metal is expensive there is an incentive to hand-craft rather than cast, especially as gold is easier to work than most other metals. The excellent properties of gold make it possible to carry out the most delicate decorative techniques. The Etruscan goldsmiths working in the seventh and sixth centuries BC in Italy created exquisite decoration with tiny spheres of gold, smaller than a modern pinhead, a technique termed granulation, and the fineness of their work has rarely been equalled since. Granules can be made by cutting pieces of gold wire and melting them in a bed of charcoal in a crucible. The surface tension causes the liquid gold to form perfect spheres. These were stuck to the goldwork with a paste of animal glue and copper salts and then heated to precisely the point when they fused to the surface. This sophisticated solid state bonding method perfected by the Etruscans and the Greeks was lost for centuries afterward and later goldsmiths struggled to produce granulation effects using solder (melted metal), a much clumsier method.

Gold fibula (bow-shaped brooch) decorated with tiny figures of lions and sphinxes, from Vulci, ancient Etruria (now in Lazio, Italy), around 675–650 BC. Details of the animals are picked out in gold granulation, and lines of granulation ornament the other parts of the brooch. L. 18.6 cm.

Wire work (filigree) is a decorative technique at which the goldsmiths of classical antiquity excelled. Fine gold wire is made today by pulling a piece of metal through graded holes of a draw-plate, each hole being slightly smaller than the last, resulting in a longer but thinner strand of gold. This technique began to be used from about AD 700 in Europe. At earlier periods fine wires were made in relatively short lengths, by tightly twisting strips of gold. The strands of wire could be given different textures for added effect and twisted together in decorative formations.

It was not until the industrialization of the mid-nineteenth century in Europe that machines were developed for making chains from fine gold links. Before this, joining each tiny loop of wire with solder (molten metal) without melting its neighbouring link in the chain was beyond the technology available and other methods had to be used to produce fine chains. One method of avoiding heat damage was to make a flexible chain by threading and folding pre-soldered loops together in a method termed loop-in-loop.

(above) One of a pair of earrings decorated with filigree, Greece, 330–300 BC. H. 6 cm

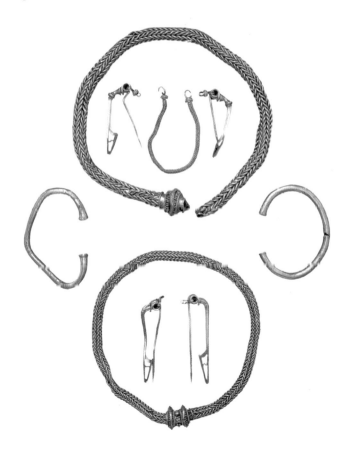

(left) Gold jewellery found near Winchester, Hampshire, dating from 75–25 BC. The necklaces are formed of flexible loop-in-loop chain more than a centimetre thick. The clasps are decorated with filigree and large granules and held shut with a split pin. L. 48 cm; Wt. 506 g.

Gold, enamel and diamond
pendant jewel, known as the
Lyte jewel, with a miniature of
James I painted by Nicholas
Hilliard, England, AD 1610.
W. 4.8 cm.

Setting stones of all types and colours in gold is a feature of goldsmiths' work through the ages. Other materials were also inlaid to create areas of colour; the techniques of melting coloured glass on to gold (enamelling), as on the Royal Gold Cup (see chapter 2), are particularly attractive. The colours of transparent enamels are enhanced by the reflection of light from the gold beneath the glass.

The historical picture which emerges of the goldsmith is far from uniform. His status in society could be anywhere from slave to influential citizen. He could be a smith of all metals, producing to the orders of the customer, or a creative artist of the highest degree. Even the finest of goldsmiths with powerful patrons were not assured of success, as is illustrated by the sad tale of Gusmin of Cologne, master goldsmith to Louis, duke of Anjou (died 1384). Gusmin 'was perfect in his works, the equal of the ancient Greek sculptors …', yet even these apparently exquisite works were melted down by the Duke when he was in need of ready finance. Gusmin retired broken-hearted to a monastery, his whole life's work reduced to bullion.

# 5 Jewellery and adornment

'*Taugte wohl des gold'nen Tandes gleißend Geschmeid,*
*auch Frauen zu shönem Schmuck?*'
'Serves as well the golden trinket's glittering dross
to deck forth a woman's grace?'
Wagner, *Das Rheingold*

The first use of gold was almost certainly for personal adornment, and gold is still central to the manufacture of jewellery. Gold jewellery is not only decorative but is always a signifier of wealth and status, and sometimes it also takes on a functional role, as in the case of dress fastenings, buckles and seal rings. Its visual appeal is obvious and it has the added advantages that it does not tarnish like silver and it is much easier to shape and join than platinum or most of the other metals that are used to make jewellery.

Gold also has the merit that it keeps its value in times of economic downturn, and it is this property which makes it a prime target for

'Master of the Animals' repoussé gold pendant from Aegina, off the south-east coast of Greece, about 1850–1550 BC. W. 6.3 cm.

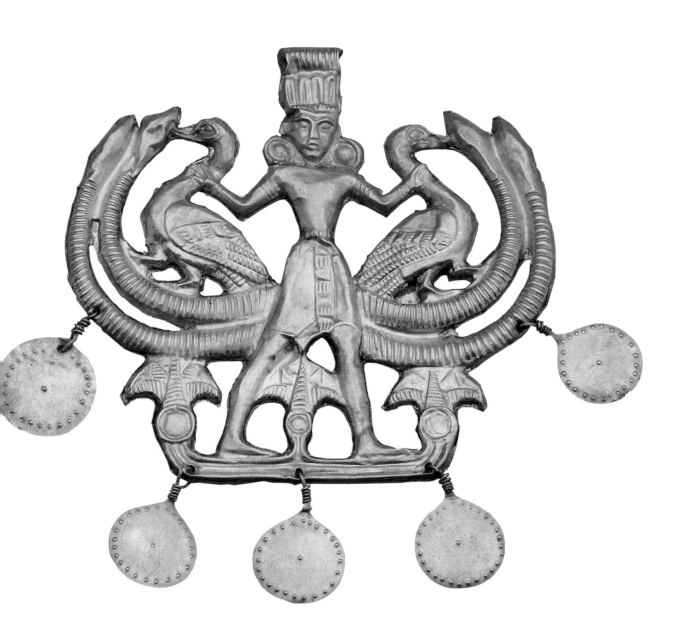

recycling – in hard times gold items are melted down for their bullion value, whatever their artistic merit. A marriage payment or dowry may consist of an agreed weight of gold, made up into jewellery to be worn by the bride, and later, when the need arises, the value of the gold can be realized by selling pieces to be melted down. This very practical use of gold jewellery as a realizable asset leaves the historical record of jewellery fashions somewhat incomplete and patchy. Even when gold jewellery escapes being melted down, there is a tendency, as fashions change, for old pieces to be altered to the current taste, and it is a matter of chance what survives to be passed down to subsequent generations.

Fragments of gold jewellery, ingots and over 400 Moroccan gold coins were found by divers on a shipwreck in Salcombe Bay, Devon, on the south coast of England. The wreck was dated by the coins to shortly after AD 1631, during the reign of Charles I, and the evidence points to the ship having sailed from Africa, ending its journey on a notoriously dangerous part of the English coast. None of the pieces of jewellery was complete and although some evidently had been inlaid with enamel and gemstones, all these materials had been deliberately removed, apparently leaving the gold ready for melting down. Element analysis showed the ingots were not pure gold but contained metals also found in the solder of the jewellery. This confirmed that the jewellery fragments were intended for melting down and that they had only been saved by the sinking of the ship. The jewellery included pendants designed for suspension from a headdress or veil, brooches and a bangle fragment, none of particularly fine workmanship but of the types which might be expected to make up the trousseau of an Islamic bride from a family of moderate means. It was only by the chance sinking of the ship that they evaded the fate of most such gold jewellery.

The Salcombe find, preserved at the bottom of the sea, is a rare survival. Most of the early jewellery which does survive has only escaped recycling because it was buried with its owner and even then

(opposite) Portrait of an Egyptian woman, encaustic on lime wood, from a mummy excavated at Hawara, Egypt, Roman period, AD 55–70. She is wearing typical jewellery of the period, gold ball earrings and a gold necklace with pendant crescent, and her robe is edged with gold. W. 21.5 cm.

such jewellery was often destroyed by grave robbers. Burials give us the clearest picture not only of what was worn but also where on the body it was worn. Thick wire bracelets were found at Varna in Bulgaria, in graves dated to the fifth millennium BC, and because these burials have been methodically excavated and recorded we know that bracelets were being worn not only on the wrists but also above the elbows. Personal adornment even at this early period included earrings of thin gold wire, necklaces of cylindrical gold beads and diadems of sheet gold. Animal-shaped plaques were pierced, perhaps for sewing on to garments.

The so-called Royal Graves at the Sumerian city of Ur, in what is now southern Iraq, dating from about 2500 BC, are a rich source of archaeological evidence. The majority of graves were robbed in antiquity but where evidence survived, the main burial was surrounded by many human bodies, both male and female. One grave had up to seventy-four such sacrificial victims. The women, wearing all their finery, followed the deceased into the tomb, ready to carry out their role in the next world. There they died, perhaps from drinking poison from cups found among the bodies. This macabre practice has been invaluable in preserving the full regalia of Sumerian court dress. Gold alternated with coloured beads of lapis lazuli and carnelian in necklaces, and large gold earrings and headdresses were also worn. The headdresses are remarkably ornate with gold leaves, flowers and ribbons cut from sheet gold.

Portraits can also be a good source of information on jewellery fashions for both men and women.

## Fastenings for clothing

Garments rarely survive in burials but the pins, brooches, buckles and buttons which fixed them together often do, telling us much about the styles of dress of the time, though only the grandest of these were made of gold. Brooches were often worn in pairs at the shoulder to secure a cloak or fix a robe, and belts were frequently used in both male and female dress to attach a purse, keys or a sword.

Even in more recent times, with the advent of buttons and zip-fasteners, dress fixings are often treated as adornments in their own right; a fine buckle draws attention to a slender waist and a sparkling brooch to a splendid décolletage.

Gold has been and still is used to adorn most parts of the body, but jewellery is primarily designed for the most visible parts: hands, arms, head and chest.

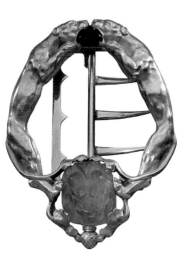

(above) Art Nouveau-style buckle made by the celebrated jewellery workshop of Frédéric Boucheron, Paris, about 1900. The glass lion's head relief was probably supplied by René Lalique. H. 7.8 cm.

(above) Three coloured gold brooch, set with diamonds. France, about 1860. This large brooch could also be worn as a hair ornament. The butterfly is on a spring which allows it to quiver with every breath and movement of the wearer. H. 9.4 cm.

(left) Gold 'dress-fastener' of the Late Bronze Age, 8th–7th century BC. These are found in Ireland and perhaps fit through double button-holes in a garment, though there are some very large examples which weigh over 1300 grams and are unlikely to have been functional dress accessories. W. 11.7 cm.

(below, left to right) Mixtec finger ring from southern Mexico, AD 1200–1521. Diam. 1.7 cm; Asante finger ring with three cannons, from Ghana, West Africa, probably 19th century. Diam. 2 cm; Burmese enamelled gold finger ring in the form of a dragon's head, set with sapphires and rubies.

# Finger rings

Finger rings are found in many cultures and rings have also been worn on thumbs and toes, or on the upper joints of the fingers, as well as in the way finger rings are more usually worn today. Rings for the upper joint were popular in the fashions of the Roman empire. This way of wearing rings is depicted in carved gravestones and they are sometimes recognizable by their small size and often slightly oval shape, which prevents them falling off too easily.

Rings are worn for their decorative value but they could also convey messages about the status of the owner. A ring may be given on betrothal and rings are exchanged on marriage. Some rings have a practical function, for example an archer's thumb ring allows the bow string to be released more sharply than is possible with the finger tips. These are less commonly made in gold but other practical rings such as signet rings for impressing the owner's mark on a wax or clay seal were frequently made of gold as it is relatively easy to engrave.

(right) Engraved gold signet ring, 14th century AD. This ring doubled in function as an archer's thumb-ring. The triangular-shaped extension on the opposite side to the engraved intaglio was worn on the palm side of the hand, to catch and draw the bowstring. The coat of arms has been identified as that of the Donati, an old Venetian family, though the ring was found at Aegium in the Greek Peloponnese. Diam. 3.8 cm.

Rings, and also lockets, were worn in memory of a deceased person, from at least as early as the Middle Ages in England.

Richard II, for example, on his death in 1400 left a gold ring to each of the nine executors of his will. By the seventeenth century it had become customary to engrave gold rings with the name and the dates of the deceased, with the decorative design on a ground of black enamel. People would leave instructions in their wills for specific sums of money to be used by the executors to buy rings for named recipients. The designs commonly show a skeleton with an hour-glass, signifying the transience of life, and sometimes a pick and shovel, a winding sheet and a skull and cross-bones with the legend 'MEMENTO MORI' ('In remembrance of death'). By the late eighteenth century the neo-classical imagery of burial urns, weeping willows and angels became popular. At this period there was a fashion for 'painted eye' rings, where a rather unnerving depiction of the eye of the deceased, often cut from an existing portrait, was set into the ring.

In the late nineteenth and early twentieth century in China, at the Manchu court of the late Qing dynasty, fine gold hair pins were worn and jewellery was even designed for the fingernails. Decorated sheaths up to 10 cm in length protected long fingernails, drawing attention to the fact that the wearer was too rich to need to carry out any manual tasks.

(above) Enamelled gold English mourning ring. The back is engraved 'Mary Dean Obt 27 Augt 1794 Aet 73' ('Mary Dean died 27 August 1794 aged 73'). Diam. 2 cm.

(left) Large gold ring, inlaid with niello, in the form of a bishop's mitre with the name of the Anglo-Saxon king Aethelwulf of Wessex. This is not believed to have been worn by the King himself but may have been a gift from him to a friend or supporter, either during his lifetime or possibly as a memento after his death. Diam. 2.8 cm.

## Adornment of the arms

Bracelets are a common form of jewellery and today are normally worn at the wrist, but this has not always been the case. In early Egypt, for example, bracelets were usually close-fitting and worn on both wrists but with matching pairs of armlets on the upper arm. In the Classical world too both the upper arm and wrists were adorned with bracelets and armlets in a variety of forms. Coiled snake bracelets and armlets have a long history of use, the coiled form allowing some flexibility in fitting closely to the contours of the arm, but bracelets of circular bands of gold were also worn, which slid over the hand like bangles. Armlets with gold and jewelled elements on a silk cord are depicted on the upper arm in portraits of the Mughal emperors and Indian princes, but bracelets in the Nasca culture of Peru were shaped and fitted more like cuffs for the wrists.

Pair of gold bracelets inscribed for the son of king Sheshonq I, founder of the 22nd Dynasty, said to be from Sais, western Nile Delta, Egypt, around 940 BC. Each bracelet is made of two segments of sheet gold, hinged together and fastened with a retractable pin and inlaid with glass. H. 4.2 cm.

(left) Indian gold and enamelled armlet, worn on the upper arm, probably from Jaipur, 18th century AD. L. (with silk cord) 39.6 cm.

(below) Roman gold bracelets, found buried in a wooden chest along with other gold jewellery, silver and gold coins and small items of silver tableware at Hoxne, Suffolk, dated 5th century AD. The bracelets include several sets of pairs. The large central armlet was worn on the upper arm. One pierced gold bracelet gives the name of the owner as Juliane. Diam (Juliane bracelet) 6.5 cm.

One of a pair of gold armlets
from the Oxus treasure.
Achaemenid Persia, about
5th–4th century BC. The
empty cells on the griffin
terminals would originally have
been inlaid with polychrome
stone, glass and faience.
H. 12.4 cm; Wt. 395.5 g.

The set of belt and sword fittings of gold and garnet work from the Anglo-Saxon ship burial at Sutton Hoo in Suffolk (early seventh century AD) is a supreme example of goldsmith's work. The grave is thought to be that of King Raedwald of East Anglia, and the grave contents certainly belonged to a man. The pair of hinged gold clasps would originally have been attached to the shoulders of a leather or fabric garment by the gold loops on the back. They are shaped to fit over the shoulders, and a pair of interlinked boars with their hip-joints in blue millefiori glass decorate the curved ends. A gold pin with a filigree animal-head terminal slides into the hinge and is kept safe by a chain. Each of the garnet and glass inlays is precisely set into gold cell walls and each is backed by a thin gold foil impressed with a design of cross-hatched fine lines which reflect the light through the inlays at every movement.

One of a pair of gold and garnet hinged clasps, from the Sutton Hoo ship burial, Suffolk, England, early 7th century AD. They would have been worn on the shoulders, attached to clothing by gold loops on the back. W. 5.4 cm.

## Ear and nose ornaments

Ears have been pierced and adorned with gold from the earliest times, and earrings were worn by men and women. The Achaemenid Persian carved reliefs on the stone walls of Persepolis, south-west Iran, dating to the sixth/fifth century BC, show the male Persian guards wearing earrings, and earrings are found in male and female graves of many cultures.

In cultures as disparate as Bronze Age Ireland, ancient Peru and modern Africa, holes in the earlobe have been deliberately stretched to take a spool-shaped plug often several centimetres in diameter. These ear plugs sometimes contain a small pebble or piece of metal to make a rattling sound when the wearer moves. The hole in the lobe can be enlarged to a remarkable size by weights suspended through the piercing and some of the Irish Bronze Age ear spools are so large they have been mistaken for round boxes.

A piercing through the septum of the nose for a ring or a sheet-gold nose ornament large enough to conceal the mouth is a feature of several pre-Columbian cultures of the Americas.

(above) Gold earring decorated with granulation, from Korea, 5th to 6th century AD. L. 6 cm.

(right) Pair of gold ear spools from a grave at Ancon, Peru. The silver alloy plug at the back would have been inserted through an enlarged hole in the earlobe. Diam. 4.2 cm.

Cast gold lime-flask, Quimbaya culture, Colombia, AD 600–1100. This seated female figure is wearing only jewellery, including an ornament attached to the septum of her nose. H. 14.5 cm.

## Ornaments for the hair, neck and chest

Decoration of the head, throat and chest is common to most cultures, though the style differs with fashion, status, the availability of gold and the forms of dress. Beads of shell, stone, bone or feathers suspended around the neck on a cord seem to be among the earliest forms of ornament, and probably the first use of gold was to make beads. At Jiskairumoko in the Lake Titicaca basin in Peru, cold-hammered native gold beads have been excavated from a 4000-year-old burial, which would appear to pre-date the use of any other metal in that area.

In both ancient Egypt and in pre-Columbian South America, for example, large gold chest ornaments or pectorals were worn mainly by men of high status (see chapter 2). The broad collars or *wesekh* of ancient Egypt, which were made of beads of brightly coloured stones and faience as well as gold, were worn by both men and women, as were torcs, the rigid metal collars of the Iron Age peoples of Europe. In ancient Greece, delicate gold necklaces and earrings with fringes of tiny pendants and diadems of gold were worn by women only.

Gold 'basket' ornaments, Boltby Scar Camp, North Yorkshire, England, Early Bronze Age. These rare finds are among the earliest goldwork from Britain. The oval or basket-shaped ornaments, with a projecting strip for attachment, are cut from gold sheet and decorated with simple punch marks. It is debatable whether they were in fact worn through the ears or wrapped around a lock or plait of hair, but they are found in male graves, positioned close to the side of the head. L. 3.1 cm.

Gold collar, found near Sintra in Portugal, about 7th century BC. A catch-plate, loosely attached by hooks, allows the collar to be taken on and off the neck. Diam. 13.1 cm.

Contemporary portrait of Sophia Schliemann wearing the delicate gold diadems, necklaces and earrings excavated by her husband, Heinrich Schliemann, from the lower levels of the ancient city of Troy (in modern Turkey).

In 1873 Heinrich Schliemann excavated the site of the remains of the ancient city of Troy at Hisarlik in what is now northern Turkey. He believed the hoard of jewellery that he discovered, along with other

metal artefacts, to be the treasure of King Priam, hidden during the siege of Troy in about 1184 BC and hence likely to have been worn by the famed beauty Helen of Troy. This exquisite jewellery consisted of gold diadems, chains, necklaces, bracelets, earrings and rings. Later excavations have shown that in fact the levels of the site where Schliemann claimed he found the treasure dates to about 2200 BC, several centuries before the Trojan Wars.

The Hellenistic period, spanning from 330 to 27 BC, was rich in jewellery made from gold from Thracian sources and the Persian gold acquired as booty by the conquests of Alexander the Great.

An unusual type of chest adornment is the body-chain which is worn over the shoulders and under the arms of the wearer, with a decorative plaque or gem setting where the chains cross over on the chest and the back. This style of ornament has a long history and can be seen in representations in both Hellenistic and Roman art, but actual examples are extremely rare. It was particularly suited to wearing

Gold diadem (detail), found on the island of Melos, Greece. Hellenistic period, 300–280 BC. The diadem is made up of three twisted gold bands with a central gold Heracles knot motif. The garnet and the delicate blue and green enamel inlays reflect the fashion for adding colour to gold jewellery L. (whole diadem) 27.9 cm.

(right) Gold body chain, said to have been found at Asyut, Egypt. Byzantine, about AD 600. L 73.4 cm.

(below) Romano-Egyptian terracotta figure, Faiyum, Egypt, 1st–2nd century AD, wearing a similar body chain. H. 19.5 cm.

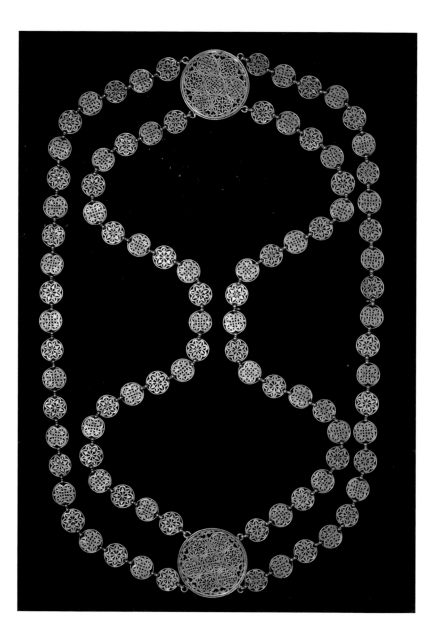

over a loose-fitting tunic of fine textile, to emphasize the female form. The example from Asyut, Egypt, would have been made either for someone of considerable girth, or perhaps it adorned a statue. Another such example, with a gem-set disc at the front and a coin on the back, was found together with other female jewellery at Hoxne, Suffolk, but that one was very much smaller and may have been made for a slender young woman.

There are still periods for which we lack the evidence from burials to truly interpret the purpose of gold artefacts which may have been worn as jewellery. It is not actually known how the so-called 'tress rings' of the late Bronze Age in Britain would have been worn, and it has been suggested that they were a form of currency, though they are generally only found singly, which would seem to preclude this idea. Although there are some parallels with Egyptian wig rings, it seems unlikely that they could have been worn in the hair. They show signs of wear on the outside but none on the inside surface, so it is unlikely they were threaded on to anything. The nearest parallels seem to be to the penannular nose rings from pre-Hispanic Colombia, which were fitted on to or through the septum of the nose, or to brass ear ornaments worn in parts of present-day Africa.

No short chapter can do justice to the many ways gold has been used for adornment of the human form. One thing is certain – all the gold jewellery that has survived the ravages of time and human greed for us to admire today amounts to only a very small proportion of what has been made and worn over the millennia.

Striped 'tress ring' (detail), from Mancetter, Warwickshire, England, Late Bronze Age. The subtle colour difference between the stripes is the result of the skilful application of a silver-rich gold strip or wire to the surface of a gold rod before it was bent into a tight curve. Diam. 1.5 cm.

# 6

# Gold, gods and death

Small gold Buddha figure, Java, Indonesia, 10th century AD. H. 6.8 cm.

'Gold is the child of Zeus, neither moth nor rust devoureth it.'
Pincar, around 522–422 BC

Places of worship were and still are adorned with gold. Solomon's Temple in Jerusalem was said to have had walls, floors and doors overlaid with gold, and gilding has decorated the domes and interiors of mosques, synagogues, temples and churches up to the modern day. Golden cult statues are recorded from early times, like the Golden Calf, worshipped by the Israelites and destroyed by Moses, a story told in Judaic, Christian and Muslim holy scriptures. In the Classical world, according to the Greek historian Thucydides, Pericles made a speech to the Athenians, then on the brink of war with Sparta, outlining the financial resources they could draw upon, including the gold and silver offerings in the temples and the forty talents of gold adorning the colossal ivory statue of Athena in the sanctuary on the Acropolis. They were required to promise to restore it again afterwards,

in equal measure or more: needless to say, golden cult statues were always at risk of destruction.

Where there was no state bank, as in the ancient Greek world, the wealth of the community could be stored under the protection of the temple, deterring thieves with fear of the vengeance of the gods. Religious centres tended to draw in wealth to themselves and inscriptions mention loans being made from temple treasuries. Suppliants wanting favours from the gods brought gifts and priests could exact payment for their services. For example, the kings of the Gupta empire of northern India (around AD 320–550) were required by the sacred Veda, which are traditional texts of the Hindu religion, to pay the temple in gold when the priests performed the horse sacrifice required to establish their kingship. The need for payment specifically in gold may have provided the impetus for the Gupta gold coin issues.

## Gifts for the gods

The gods, like human rulers, were expected to be pleased by gifts and tribute. Financing the building of a beautiful place of worship has been the chosen route to redemption for many a wealthy patron, no doubt to make up for ethically dubious behaviour in the accumulation of that wealth. The temple of Artemis at Ephesus, a prosperous Greek city on the west coast of Asia Minor, rebuilt on the ruins of an earlier temple, was numbered among the Seven Wonders of the World, although nothing of it stands today. This huge building project was funded by King Croesus of Lydia and in the foundation deposit were found some of the earliest electrum (gold-silver) coins. The cult of Artemis was associated with childbirth, and with wild animals: early representations show her in the pose of 'Mistress of Animals', with a wild creature at each hand; later Roman depictions give her a bow and arrow, as Diana the huntress.

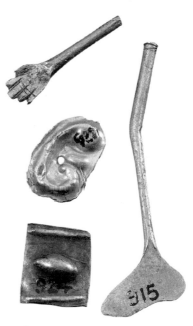

(below) Gold miniature models of parts of the body, from the Greek temple of Artemis at Ephesus (modern Turkey), around 650–600 BC. Pilgrims offered them in hope of, or in gratitude for, a cure from an illness. L (leg and foot). 2.7 cm.

The temple housed the great cult image of Artemis, drawing pilgrims from hundreds of miles around who wished to petition the goddess for cures for illness, good fortune in business deals or for the birth of a child. Pilgrims have left offerings to their gods at sacred sites all around the world. These were not necessarily wealthy people, but they gave the best they could afford in the hope of reward.

(opposite) Gold plaques from Ashwell, Hertfordshire, England, dated to the Roman occupation of Britain. The plaques are inscribed with the name of a previously unknown Celtic goddess, SENUNA, who is portrayed as resembling the Roman goddess Minerva. There are projections from the bases of the thin gold plaques which might have been pushed into soft ground to stand them upright. The inscriptions name particular suppliants and refer to the fulfilment of their vows to the goddess. H. 6–16 cm.

(left) Gold plaque from the Oxus treasure, probably from a temple in the region of Takht-i Kuwad, modern Tadjikistan. Achaemenid, 5th–4th century BC. The figure is thought to be a priest because he is carrying a bundle of sticks known as a barsom, which were originally grasses distributed during religious ceremonies. This is one of about fifty plaques, mostly depicting people, which may have been gifts to the temple from worshippers. H. 15 cm.

(opposite top right) Gold funerary mask from a burial ground on the site of the former Late Assyrian citadel of Nineveh, northern Iraq, 2nd century AD. Two bodies were found in the grave, one identified as a woman. Both had gold masks and eye coverings. Finger rings, gold buttons and beads and an already antique coin of the Roman emperor Tiberius (reigned AD 14–37) had been placed with the bodies. L. 13.9 cm.

(opposite below left) Sheet gold headdress with repoussé decoration, worn by a mummy from the Camaná valley, Peru, 1st century AD. The large face motif in the centre was probably that of a rain god, with the lines beneath the eyes representing tears. L. 23 cm.

(opposite below right) Gilded and painted cartonnage mummy mask, Egypt, late 1st century BC/early 1st century AD. Such masks were worn over the wrapped head of the mummified body. H. 44 cm.

(right) Purse lid and some of the gold coins and ingots found within it. From the ship burial at Sutton Hoo, Suffolk, England, early 7th century AD. W. (purse frame) 19 cm.

## Gold for the dead

Belief in life after death is a fundamental tenet of most religions. Some also believed that it was possible and indeed necessary to equip the dead with worldly items for use in the next life. At Sutton Hoo a man believed to be the Anglo-Saxon ruler of East Anglia, Raedwald (599–624/5), was buried in a wooden long ship about 27 metres long, under a large earth mound along with gold regalia and all that would be needed for a royal feast. Among these burial goods there was a purse that contained 37 gold tremisses, each from a different Frankish mint, three blank coins and two small gold ingots. The blanks bring the number of 'coins' up to 40, perhaps the amount needed to pay the men who would row the ship into the 'Otherworld', while the ingots were to pay two steersmen.

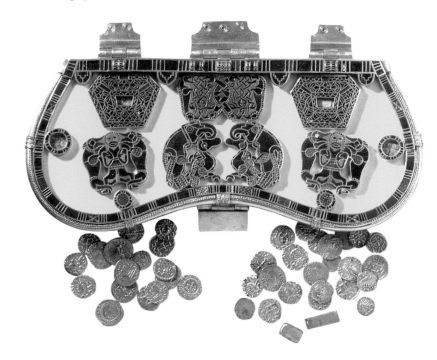

From the emperors of China to the pharaohs of Egypt and from the lords of Sicán (Peru) to the warrior aristocracy of Thrace (the Balkans), the powerful and the wealthy went to their graves with everything appropriate to their status in life. Some of these preparations for burial took most of a lifetime and removed from circulation significant quantities of gold and other precious commodities. Even those of modest means would be buried with their possessions to ensure a comfortable afterlife. These widespread beliefs ensured that grave-robbing became a worldwide occupation.

Album leaf of Shah Jahan and his sons visiting a saint. Drawn and gilded on paper. Early 18th century AD, Islamic, Mughal style.

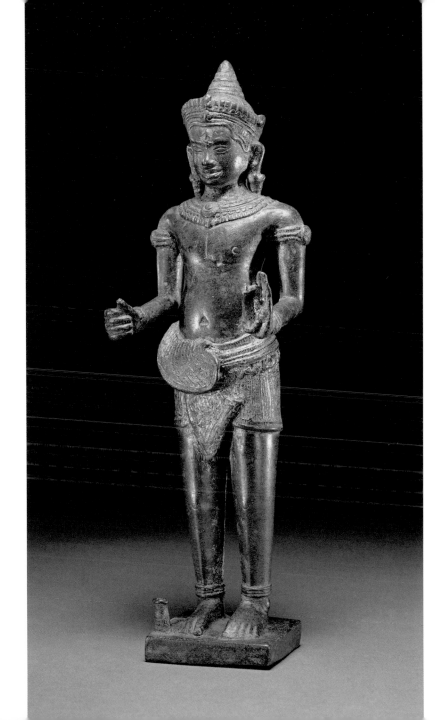

Gilt bronze figure representing the Hindu deity Shiva, from Cambodia, 11th century AD. H. 28 cm.

## Reliquaries

Relics are important to the followers of some forms of Buddhism, Christianity, Hinduism, Islam, shamanism and many other religions. Holy relics, items associated with saints or with figures such as Christ, the Buddha or Muhammad were not mere keepsakes. They were, and still are, attributed with healing powers and associated with miraculous events. Reliquaries are the containers that store and display relics. Since the relics themselves were considered 'more valuable than precious stones and more to be esteemed than gold', they were often enshrined in containers crafted of, or covered by, gold. These precious objects were a major art form across Europe and Byzantium throughout the Middle Ages. Pious patrons donated gold and gems to adorn the relics, though these reliquaries were not always as sumptuous as they might appear. The life-sized gold and gem-encrusted reliquary head of St Eustace is in fact only gold-plated and many of the 'gems' are glass. According to legend, St Eustace was a general under the Roman emperor Trajan (reigned AD 98–117). He was converted to Christianity while hunting, after seeing a vision of a stag with a luminous crucifix between its antlers. Some time later, after victory in battle, he refused to join in thanksgiving to the Roman gods and was burnt to death with his wife and sons.

(opposite) St Eustace reliquary head from Basle, Switzerland, around AD 1200. This head was made to house fragments of the skull of the saint. H. 33.5 cm.

(left) Gold and enamel reliquary cross said to have been found on the site of the Great Palace at Constantinople, early 11th century AD. The cross hinges at the top and bottom; the cap on one end of the top hinge unscrews so the two sides can be opened to reveal a relic, possibly a fragment from the True Cross. W. 3 cm.

The cult of relic worship was very important in early Buddhism. Large mounds (stupas) were constructed to house relics of the Buddha or his monks. Buddhist relics were commonly buried with 'seven precious things', items of gold being included in this category. A reliquary bearing an inscription to the effect that it contained bones of the Buddha was found in the nineteenth century at Bimaran, Gandhara (in modern Afghanistan), but the lid of the reliquary and

Bimaran reliquary of gold set with garnets, from stupa 2 at Bimaran, Gandhara (modern Afghanistan), around AD 50. The frieze bears one of the earliest depictions of the Buddha. H. 6.7 cm.

the bones were missing. Small pearls, beads of precious and semi-precious stones and four coins were buried with it.

## Cults of the sun

The association of gold with the energy of the sun is found in many mythologies. Among Andean cultures for example, gold was called 'sweat of the sun' and silver 'tears of the moon'. Gold was the attribute of the Inca sun god Inti and the Inca temple of the sun at Cusco, Peru, was reputed to have had a garden filled with naturalistic models of plants, animals and people, all cast in gold and silver. The Muisca of Colombia built temples to the Sun and Moon and had a complex belief system with at least four principal deities and many lesser spiritual entities. Votive offerings were important to religious observance; numbers of intricately detailed miniature figures of people, animals and weapons etc. cast in gold, tumbaga (copper-gold alloy) and copper were placed at locations such as lakes.

Gold was considered by the ancient Egyptians to be a divine and indestructible metal and their sun god Ra had flesh of pure gold. The skin of Egyptian deities was believed to be golden and the sister goddesses Isis and Nephthys were depicted on the ends of sarcophagi kneeling on the hieroglyph meaning 'gold'. In more recent times, gold was considered by the Asante of west Africa to be the earthly counterpart to the sun and the physical manifestation of life's vital force (kra). Gold in the ruler's regalia represented his purity and vigour and the king's Golden Stool is the most important symbol at the Asante court. The traditional stories of the Asante tell that in AD 1701, as Osei Tutu was sitting beneath a tree, the Golden Stool descended from the sky and came to rest in his lap, confirming divine support for his rule.

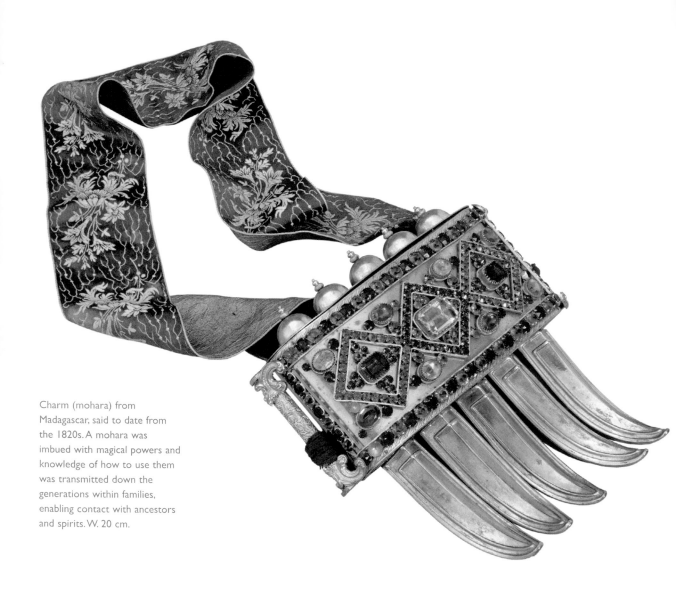

Charm (mohara) from
Madagascar, said to date from
the 1820s. A mohara was
imbued with magical powers and
knowledge of how to use them
was transmitted down the
generations within families,
enabling contact with ancestors
and spirits. W. 20 cm.

Girdle prayer book, London, England, AD 1540–45. This miniature Christian prayer book enclosed within enamelled gold covers was worn hanging from a belt at the waist and was particularly fashionable for ladies of rank in the English court between around 1530 and 1560. H. 6.35 cm.

## Gold and godliness

From all that has said before it might appear that gold has always been a well-integrated feature of religious observances, but this is by no means true. In many religions, strict observance requires abstinence from worldly things, and possession, use and wearing of gold certainly fall into the category of worldly commodities. From the earliest days of Islam, for example, the role of precious metal and money caused deep unease but money was essential for the smooth running of many aspects of society, from trade to taxation. To control this perceived bad influence, rules were established which laid down the proportion of wealth that should be given as alms and also forbade usury (riba), which was explicitly banned in the holy writings of the Qur'an. There

Obverse of a dinar minted in Qumm, Iran, AH 337 (AD 967), with the proclamation of the Islamic faith (shahada) in Kufic script: 'There is no God except Allah. He is alone. [There is] no partner to Him. He is alone'. Instead of a portrait of the Buwahid ruler, his title appears in the centre, with the mint and the date in the legend around the margin. Diam. 2.1 cm.

were prohibitions on the use of precious metal utensils and on the wearing of gold by men, but gold coinage was seen as a necessity, though a religious text generally replaced the conventional portrait of the issuer on the coin.

Many religions make deliberate provisions to reconcile spiritual and worldly wealth. For example, Jain monks and nuns, followers of Mahavira, the Buddha's contemporary, tread a path of severe self-denial, but lay followers are encouraged to follow disciplined business and professional activity, taking leading roles in India's financial systems. Gold has always held an ambiguous position in world religions.

Throughout history gold can be seen to have played a part in the affirmation of a ruler's divine right to kingship; it also underpins the wealth and power of religious establishments and it is used to embellish places of worship and hence honour the deity. Nevertheless it is also seen as the root of all evil.

# 7 All that glitters

'Let Art learn so much alchemy that it tinctures all in gold.'
Jean de Meung, *Roman de la Rose* (1277)

Leaf gilding on an illuminated medieval manuscript depicting Avarice, one of the Seven Deadly Sins. Three demons encourage the miser as he counts his gold.

The purity of gold is not always what it seems and many ingenious methods have been devised to produce fraudulent imitations. Most gold naturally contains some silver, typically between 5% and 30% by weight. Copper is the only other metal which occurs above trace levels in natural gold, though it is rarely found in quantities greater than 2%. Exactly what was understood as 'gold' by the ancients is not known, for they did not have the modern scientific concept of pure elements. The early writings of Mesopotamia tell of gold from different regions as if they were different metals. The characteristics that they must have perceived would have been physical properties such as colour, hardness and the amount of heat needed to melt them. All of these properties are affected by the proportions of silver and copper in a gold alloy (mixture of metals), and in some cultures these properties, especially

colour, have been valued more than purity of gold. In the sixteenth century, the Taíno of the Caribbean valued guanin, an alloy of copper-gold-silver which had copper as its main component, for its colour and smell. Because of the lack of availability of copper in the insular Caribbean, they were not able to make this alloy and they would trade many times the weight of pure gold in exchange for it.

## Gold refining in antiquity

From at least the third millennium BC, the removal of the copper alloyed with gold was being carried out, but there seems to have been

Musicians on a pendant from Costa Rica. It was cast in a single piece from a multi-component wax model. The metal is an alloy of copper and gold in almost equal quantities. Copper was deliberately leached from the surface leaving a thin layer enriched in gold, a method known as depletion gilding.

no attempt to remove the silver that natural gold almost always contains. The concept of removing silver as well as copper may have begun with the wish to change the surface colour of artefacts to a brighter gold by removing both copper and silver from the surface, leaving a gold-rich skin. This process, known as depletion gilding, was certainly known as early as the third millennium BC at Ur in Mesopotamia, and it is well known from the guanin or tumbaga (copper-rich gold) artefacts of early civilizations in Central and South America. The active agents thought to have been used alone or in combination in antiquity to remove copper and silver are sodium chloride (common salt), alum, ferric sulphate and potassium nitrate (nitre, also known as saltpetre).

The advent of precious metal coinage was a natural trigger for the process of refining gold, as it introduced the need for a trusted standard of metal fineness. Copper can be removed by a process known as cupellation, whereby lead is melted with the impure metal and the resulting mixture is subjected to a blast of air at about 1100°C. This oxidizes the lead and copper but leaves the two noble metals, gold and silver, unaffected.

Before the discovery of mineral acids in the Middle Ages, the only method of parting gold and silver was a solid state process known as cementation (from the so-called 'cement' – the mixture of active agents and carrier used in the reaction), which employed similar active ingredients to those used for depletion gilding. The impure gold, either as fine dust or beaten very thin, is packed with the active agent, for example salt, into a ceramic container called a parting vessel and heated for many hours at a temperature below the melting point of the impure gold. The salt vapours attack the silver, converting it to volatile silver chloride but leaving behind pure gold. The parting vessel is usually lined with an inert carrier such as clay or brick dust which serves to absorb the silver chloride, and pure silver can then be recovered by a further process of cupellation. In the Middle Ages sulphur parting, using either elemental sulphur or, more usually, an iron or antimony sulphide, was introduced.

## Assay – the measurement of gold purity

The ability to determine the purity of gold is fundamental to any refining process. Simply melting gold in an open container will cause oxidation of copper in the alloy and the resulting loss of weight can be measured. There are records from ancient Egypt and Mesopotamia of weight loss from gold alloys which were 'put to the fire' but this would not have given any estimate of the silver content without a further cementation stage. Simple methods to test for counterfeit gold coins, such as biting to test hardness, and comparing the ringing sound when dropped on a hard surface, were only broad indicators. The best tool available, certainly from classical antiquity and continuing in use to the modern day, is the touchstone. A touchstone is a piece of smooth, dark, fine-grained stone on which the gold to be tested is rubbed, leaving a streak of metal on the stone. The colour of this streak can be compared to a streak from a piece of gold of known composition. The problem with this method is that it only works for binary alloys, that is gold-silver alloy or gold-copper alloy, but not where all three metals are present, which is usually the case for gold coins, jewellery and plate.

Gold items are seldom made of pure gold. Most gold jewellery today is made from refined gold which has then been alloyed with silver and copper. Copper increases the hardness of the metal, making it more hard-wearing, and the addition of silver balances the red colour of the added copper, while keeping good malleability. The purity of gold is now described in terms of carats, or karats in the United States. In Byzantium in the fourth century AD the carat was adopted as a measure of gold purity, equivalent to one twenty-fourth part of the weight of the gold solidus coin of Constantine I. Thus pure gold is 24 carat, 18 carat is 75 per cent gold, and 9 carat contains only 37 per cent gold. The carat as a unit of measurement for gold spread until it is now almost universally used, though there has been no one accepted standard for gold jewellery and plate and standards differ from country to country today.

Hallmarks on the base of a gold beaker. Top left is the leopard's head crowned, signifying the London Assay Office, beside the London date letter 'h' for 1685. Below are the maker's mark GG (George Garthorne) and the lion passant mark for 22-carat gold.

## Hallmarks

Marks stamped on gold and silver, often called hallmarks, were introduced to give the customer assurance of the quality they were paying for and to exclude from the trade goldsmiths who were undercutting the prices of genuine quality gold with falsely described jewellery and plate. Regulations to control standards were drawn up in Paris around 1268 and other European cities soon followed suit. In England, a statute of Edward I dating to 1300 refers to gold and silver standards, and officials travelled around checking goldsmiths' shops for compliance. This was followed by the first royal charter of the Goldsmiths' Company in 1327, marking its formal foundation in London as a craft guild. The maker's mark, the unique mark denoting a particular goldsmith's workshop, was first introduced in 1363, but it was not until 1478 that the Assay Office at Goldsmiths' Hall in London was established and all makers were required to bring their wares there to be tested for gold purity and to receive the hallmark. A letter of the alphabet denoting the year the assay was carried out was also marked on to each piece.

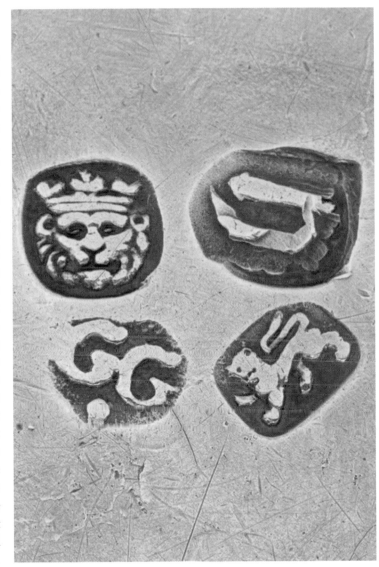

## Coloured gold

Multi-coloured effects and inlays of different alloys and materials were popular in antiquity, as was the manipulation of the colour of gold by alloying. Attractive shades can be produced by alloying pure gold with different proportions of other metals. For example, copper alloyed with gold gives it a pink tone, whereas adding silver will give a greenish tinge to gold. In more recent times, experimentation in mixing other metals with gold has produced interesting metal colours. For example, 25% iron in gold gives a blue tone, adding about 15% iron turns the gold grey, and aluminium gives a purple colour. White gold can be produced by alloying with silver, palladium or the much cheaper zinc and nickel. Another approach is to treat the metal surface to produce a thin coloured layer, known as a patina. One specialized alloy of a few per cent gold in copper was highly valued in antiquity for its property of taking a beautiful matt black patina, which contrasts well with gold and silver inlays. The Romans (opposite) Plaque of the black patinated alloy known as Corinthian bronze, Roman, 1st century AD. It is inlaid with silver and gold cupids. H. 7.1 cm.

(below) Inlaid sword guard (tsuba) made of shakudo, the Japanese black patinated alloy of copper with a few per cent gold, 19th century. W. 6.1 cm.

(above) Three-coloured gold brooch of a dove on a fruiting peach spray, probably English, around 1850. The yellow body of the dove is 100% gold, the greenish leaf and peach have 23% silver in the gold alloy, and the pink patch on the peach has about 30% copper. W. 5.3 cm.

Carillon clock with a gilded brass case. This weight-driven musical clock was made in Strasbourg in 1589 by Isaac Habrecht. H. 140 cm.

called it Corinthian bronze, the Japanese call it shakudo, and in Egypt, where it was used for fine statuettes of pharaohs and other polychrome inlaid items, it appears under the name of *Hmpty km*.

## Gilding of metals and other materials

Gilding, that is applying a thin layer of gold to the surface of metal and non-metallic items, is an economical method of producing an attractive golden finish. The earliest method of gilding metal was by mechanically attaching gold foil, usually by hammering the foil into grooves cut into copper or bronze or folding it around the edges of the object. With the ability to refine gold came the possibility to hammer the soft pure gold much thinner. Small sheets of gold, so thin that they cannot support their own weight, are known as gold leaf. Gold leaf has been used not only to gild metal but to gild wood, paper, parchment, plaster and many other materials.

Mercury gilding, sometimes also known as fire-gilding, is known first from China in the Warring States period (468–221 BC). Among the civilizations of the Mediterranean it became widespread in the second century AD, but rare examples of the method are found in Europe around the first century BC. The advantage of mercury gilding is that it can be used to produce a thin continuous layer of gold, with no need for joins or adhesives. Gold forms a liquid amalgam with mercury which can be applied very successfully to silver and copper alloys. The amalgam layer is heated to drive off surplus mercury, and the resulting yellow surface is burnished until it is smooth and shiny. Mercury compounds and vapours are highly poisonous, and it was only with the widespread use of electro-plating in the mid-nineteenth century that it was finally replaced by a less toxic process. Electro-gilding, as its name suggests, utilizes an electrical current to deposit gold on to items immersed in a gold solution; it is a fast and efficient method of gilding many pieces at once.

(left) Mamluk goblet of gilded and enamelled glass decorated with fish and eels, from Cairo, Egypt, AD 1300–1320. H. 11 cm.

(below) Silk textile (detail) with gold threads made by winding a thin strip of metal foil around a strand of silk. Indonesia, 20th century.

(opposite) Bowl of sandwich
gold glass from a tomb in
Canosa, Puglia, southern Italy.
Hellenistic, around 250 BC. The
floral design in gold leaf is
sandwiched between two layers
of clear glass. Diam. 20 cm.

(right) One of a pair of gilded
Sèvres porcelain ice pails, made
in 1811 and presented to the
Emperor Francis II of Austria by
Napoleon. H 34.2 cm.

(above) Gold-plated lead ring with an intaglio of a female figure, probably intended to be accepted as solid gold. Roman, 1st–3rd century AD. Diam. 2.3 cm; Wt. 16 g.

(below) 20th-century forgery of a pre-Columbian gold object. H. 5 cm.

It is of course not just other metals which can be gilded. Liquid gold, a suspension of powdered gold, is used to decorate porcelain and glass. It was first manufactured by the Royal Porcelain Factory at Meissen in Saxony in the early 1830s. The liquid gold is applied by brush, screen printing or spraying and then heated to burn off the organic components, leaving a thin layer of 22-carat gold.

## Fakes and forgeries

Anything that is highly valued is likely to be copied in cheaper materials. Gilding has often been used not just to enhance the appearance of an object, but also to create the impression that it is solid gold, as in the case of this Roman intaglio ring. Lead is closer to the weight of gold than silver or copper and its use here must indicate the deliberate intention of the Roman goldsmith to deceive the recipient that it was solid gold.

Even the very first gold coinage was almost immediately copied by gold-plated forgeries. The forgery of coins was such a threat to economic stability that it was punishable by death or horrible mutilation in medieval England. With the increasing interest in collecting antiquities, these too have become a focus for forgers, whether made of gold or any other material. The fragmentary ornament in the style of Chimu culture artefacts from Peru is imitating gold but is actually made of electroformed copper, a method using an electrical current to deposit metal from solution on to a mould. This technique of manufacture was not of course available to the pre-Columbian goldsmith. It was electro-gilded to give a golden surface.

(above) Genuine gold-plated spectacles, Ivory Coast, West Africa, 20th century, as worn by a modern-day chief. W. 12.3 cm.

## Enduring appeal

Arguably gold has had more power over the human mind than any other metal, from its first appearance in prehistory in the graves of high-status individuals to modern-day 'bling' worn for public display. Gold has provoked admiration and envy in equal measure throughout history. It is trusted to hold its value in times of financial crisis yet distrusted as a corrupter of moral values. The properties of the metal, particularly its appearance and resistance to corrosion, have been integral to its success in all the aspects outlined in the previous chapters, and even today the properties of gold are being exploited in new ways. It makes durable fillings and crowns in dentistry, the gold isotope Au-198 is used in some cancer treatments and for treating other diseases, its electrical conductivity gives it applications in electronics and its ability to reflect infra-red and visible light was exploited for protective coatings in the US Apollo space programme. A world without gold would indeed be a poorer place.

(opposite) Mercury-gilded contemporary forgery of a gold noble of Henry VI, who reigned AD 1422–61.

# Suggestions for further reading

## General

Grimwade, M., *Introduction to Precious Metals*. Butterworth (Newnes Technical Books), London 1985

Herrington, R., Stanley, C. and Symes, R., *Gold*. Natural History Museum, London 1999

Michaelson, C., *Gilded Dragons: Buried treasures from China's golden ages*. British Museum Press, London 1999

Morteani, G. and Northover, J.P. (eds), *Prehistoric Gold in Europe: Mines, metallurgy and manufacture*. Kluwer Academic Publishers, London 1995

Ogden, J., 'Aesthetic and technical considerations regarding the colour and texture of ancient goldwork', in S. La Niece and P. Craddock (eds), *Metal Plating and Patination*. Butterworth Heinemann, London 1993, 39–49

Rose, T.K., *The Metallurgy of Gold*. Charles Griffin and Company Ltd, London 1915

Sutherland, C.H.V., *Gold*. Thames and Hudson, London 1959

## Gold testing, refining and coinage

Forbes, J.S., *Hallmark: A history of the Assay Office*. The Goldsmiths' Company and Unicorn Press, London 1999

Moore, D.T. and Oddy, W.A., 'Touchstones: some aspects of their nomenclature, petrography and provenance', *Journal of Archaeological Science* 12, 1985

Ramage, A. and Craddock, P., *King Croesus' Gold: Excavations at Sardis and the history of gold refining*. British Museum Press, London 2000

Eagleton, C. and Williams, J. (eds), *Money: A History*. 2nd edn, British Museum Press, London 2007

## Goldsmiths and their techniques

Campbell, M., 'Gold, silver and precious stones', in J. Blair and N. Ramsay (eds), *English Medieval Industries*. Hambledon Press, London 1991

Cherry, J., *Medieval Craftsmen: Goldsmiths*. British Museum Press, London 1992

Coatsworth, E. and Pinder, M., *The Art of the Anglo-Saxon Goldsmith*. Boydell Press, Woodbridge, Suffolk 2002

Craddock, P. and Guimlia-Mair, A., 'Hsmn-km, Corinthian bronze and shakudo', in S. La Niece and P. Craddock (eds), *Metal Plating and Patination*. Butterworth Heinemann, London 1993, 101–27

Drayman-Weisser, T. (ed.), *Gilded Metals: History, technology and conservation* (AIC conference proceedings). Archetype, London 2000

La Niece, S., 'Niello before the Romans', *Jewellery Studies* 8, 1998, 49–56

Lightbown, R.W., *Secular Goldsmiths' Work in Medieval France*. London 1978

Nicolini, G., *Techniques des Ors Antiques: la Bijouterie Ibérique du VII au IV Siècle*. Picard 1990

Perea, A., Montero, I. and Garcia-Vuelta, O. (eds), *Tecnología del Oro Antiquo: Europa y América*. CSIC, Madrid 2004

Ogden, J.M., *Age and Authenticity: The materials and techniques of the 18th and 19th century goldsmiths*. National Association of Goldsmiths, London 1999

Theophilus, *On Divers Arts*. Trans. J.G. Hawthorne and C.S. Smith, Dover 1979

Ward, R., *Islamic Metalwork*. London 1993

Williams, D. and Ogden, J.M. (eds), *The Art of the Greek Goldsmith*. British Museum Press, London 1998

Wolters, J., *Die Granulation. Geschichte und Technik einer alten Goldschmiedekunst*. Callwey Verlag München 1983

## Adornment

Aldred, C., *Jewels of the Pharaohs*. London 1971

Andrews, C., *Ancient Egyptian Jewellery*. British Museum Press, London 1997

Bray, W., *The Gold of Eldorado*. Royal Academy, London 1978

Bury, S., *Jewellery 1789–1910*. 2 vols, Woodbridge, Suffolk 1991

Evans J., *A History of Jewellery 1100–1870*. London 1970

Fisher, A., *Africa Adorned*. London 1984

Gere, C., *European and American Jewellery 1830–1914*. London 1975

Johns, C., *The Jewellery of Roman Britain*. UCL Press, London 1996

McEwan, C. (ed.), *Precolumbian Gold: Technology, style and iconography*. British Museum Press, London 2000

Ogden, J.M., *Jewellery of the Ancient World*. Trefoil, London 1994

Phillips, C., *Jewelry from Antiquity to the Present*. Thames and Hudson, London 1996

Scarisbrick, D., *Jewellery in Britain 1066–1837*. Norwich 1994

Tait, H. (ed.), *Seven Thousand Years of Jewellery*. British Museum Press, London 1986

Taylor, J.J., *Bronze Age Gold Work of the British Isles*. Cambridge University Press 1980

Untracht, O., *Jewelry Concepts and Technology*. London 1982

Untracht, O., *Traditional Jewelry of India*. London 1997

Williams, D. and Ogden, J., *Greek Gold: Jewellery of the classical world*. British Museum Press, London 1994

## Journals

*Aurum*
*Gold Bulletin*
*Jewellery Studies*
*Boletin*, Museo del Oro, Bogota, Colombia

## Websites

www.britishmuseum.org
www.thegoldsmiths.co.uk
www.societyofjewelleryhistorians.ac.uk

## Illustration acknowledgements

Illustrations are © The Trustees of the British Museum except where otherwise noted.

*page*

2 GR 1908,0414.1
4 ME 1897,1231.23 (Bequeathed by Sir A.W. Franks)
6 ME 1897,1231.18 (Bequeathed by Sir A.W. Franks)
9 GR 1892,0520.1
10 ME 1928,1010.7
11 (top) P&E T.0.a (Bequeathed by Charles Townley) (below) ME 1949,0212.7
12 Natural History Museum 30884
13 © P.T. Craddock
14 Powerhouse Museum, Sydney, photo Penelope Clay
15 P&E 1913,1220.35 (Bequeathed by the Revd A.H.S. Barwell)
16 (top) AOA Am1952,10.1 (below) AES 1872,0604.1098 & 9
17 GR 1946,0702.2 (Given by Mrs M.W. Acworth)
18 AOA Am1896,0203.1
19 GR 1894,1204.2
20 P&E 1974,1201.342–6 (Given by the Pitt-Rivers Museum, Oxford)
21 (left) P&E WB.152 (Bequeathed by Baron Ferdinand de Rothschild) (right) ME 1972,0617.1
23 ME 1928,1010.6
24 AES 1923,1013.2
25 ASIA 1937,0416.218
26 ASIA 1949,1213.1–2
27 (left) GR 1872,0604.842 (right) ME 1929,1017.3
28 (top) AOA Am,+.7834 (Christy Fund); (below) GR 1898,1201.185
29 (left) ME 1974,0617,0.21.3; (right) Crown Jewels of England © AKG-Images
30 P&E 1836,0902.1 (Given by the Revd G. Rushleigh)
31 ME 1897,1231.7 (Bequeathed by Sir A.W. Franks)

32 P&E 1951,0402.2 (Purchased with the assistance of the Art Fund)
33 GR 1869,1025.3 (Given by Henry Danby Seymour)
34 P&E 1892,0501.1 (Acquired with contributions from HM Treasury, the Worshipful Company of Goldsmiths, Sir A.W. Franks and others)
35 AOA Am,+.342
36 CM 1999,1207.426
37 AOA Af1900,0427.44
39 ME 1948,1009,0.69 (Bequeathed by P.C. Manuk and Miss G.M. Coles through the Art Fund)
40 AES 1974,0223.1
41 CM RPK,p146B.1
43 GR B.11528–30, GR 1894,1207.2
44 (left) CM 1883,0516.1 (right) P&E 1859,0512.1
45 CM 1883,0506.1
46 CM 1922,0424.64
47 CM 1996,0111.1–100
48 CM 1947,0604.3 (Given by E.T. Sykes)
49 CM 1980,1136.1
50 CM 1935,0401.8239 (Bequeathed by T.B. Clarke-Thornhill)
51 CM SSB,168.66 (Given by Dorothea Banks)
52 CM 1855,0321.14
53 CM 1862,0925.1
54 CM 1967,1208.1–9
56 P&E 2005,0604.1
57 P&E 2005,0604.1–2
59 PD 1893,0411.10.2
60 P&E 1986,0403.1
61 © Marc Quinn
62 AES 1869,1025.2 (Given by Henry Danby Seymour)
63 PD 1886,0111.84
64 ME 1927,0525.1
65 (top) P&E 1980,0501.1; (below) AOA Am1904,1031.1
66 (top) P&E 2003,0501.1 (Purchased with the assistance of the Heritage Lottery Fund, the Art Fund and the British Museum Friends); (below) AOA Am1921,0721.1
67 P&E 1997,0707.1 (Purchased with the assistance of the Heritage Lottery Fund and the British Museum Friends)

68 GR 1862,0512.16
69 (left) P&E 2001,0901.1–10 (Purchased with the assistance of the National Heritage Memorial Fund, the Art Fund and the British Museum Friends); (right) GR 1877,0901.16
70 P&E WB.167 (Bequeathed by Baron Ferdinand de Rothschild)
71 ME 1964,0212.2 (Given by Ralph Pinder-Wilson)
73 GR 1892,0520.8
75 AES 1994,0521.11
76 P&E 1939,1010.1 (Given by Mrs Pretty)
77 (top left) P&E 1978,1002.645, (top right) P&E 1978,1002.956 (both given by Professor and Mrs Hull Grundy); (below) P&E 1834,1222.1
78 (top, l–r) ASIA AF.2373 (Bequeathed by Sir A.W. Franks), AOA Af1979,01.4661, AOA Am1914,0328.1(Given by Miss Thornton); (below) P&E AF.568 (Bequeathed by Sir A.W. Franks)
79 (left) P&E 1829,1114.1 (Given by the Earl of Radnor); (right) P&E AF.1732 (Bequeathed by Sir A.W. Franks)
80 AES 1850,0817.1–2
81 (top) ASIA 1961,1016.3 (Bequeathed by Louis Clarke); (below) P&E 1994,0408.11–29 (Purchased with the assistance of the National Heritage Memorial Fund, the Art Fund and the British Museum Friends)
82 ME 1897,1231.116 (Bequeathed by Sir A.W. Franks)
83 P&E 1939,1010.4–5 (Given by Mrs Pretty)
84 (left) ASIA 1938,0524.238; (right) AOA Am1920,1013.2a–b (Given by G. Lockett)
85 AOA Am1940,11.2 (Given by The Art Fund)
86 P&E 1940,0404.1 (Given by Sir C. Smith-Dodsworth)
87 P&E 1900,0727.1
88 © AKG-Images
89 GR 1872,0604.815
90 (left) GR 1926,0930.42;

(right) P&E 1916,0704.1 (Given by Mrs Burns)
91 P&E 1848,0217.1
93 ASIA 1888,1110.53
95 (left) GR 1861,0425.3; (right) GR 1907,1201.19,21–23
96 P&E 2003,0901.8–14 (Purchased with the assistance of the Art Fund and the British Museum Friends)
97 ME 1897,1231.48 (Bequeathed by Sir A.W. Franks)
98 P&E 1939,1010.3 (Given by Mrs Pretty); CM 1939,1003.3,13,27–41
99 (left) AOA Am,+.7819; (top right) ME 1856,0909.67; (below right) AES 1897,0511.188
100 ME 1920,0917,0.35
101 ASIA 1960,1114.1
102 P&E 1850,1127.1
103 P&E 1965,0604.1
104 ASIA 1900,0209.1
106 AOA Af1900,0524.34a–b
107 P&E 1894,0729.1 (Given by Sir A.W. Franks)
108 CM 1936,0605.5
109 ASIA 1880.29
111 British Library, MS 70560 Add.28162, fol. 9v
112 AOA Am1907,0618.1
115 P&E 2007,8037.1 (Accepted by HM Government in lieu of Inheritance Tax)
116 GR 1979,1231.1
117 (left) ASIA 1952,0211.28 (Given by Mrs Agnes F. Barden) (right) P&E 1978,1002.928 (Given by Professor and Mrs Hull Grundy)
118 P&E 1888,1201.100 (Bequeathed by Octavius Morgan)
119 (left) ME 1879,0522.68; (right) ASIA As1955,06.2
120 P&E 1985,1203.1
121 P&E 1871,0518.2 (Given by Felix Slade)
122 (top) GR 1772,0314.6 (below left) AOA Am1947,21.1.a–c (Given by Mr & Mrs W.T. Smithies) (below right) CM 1968,0412.1132
123 AOA Af1987,03.1

# Index